Rapid Detection of Infectious Agents

INFECTIOUS AGENTS AND PATHOGENESIS

Series Editors: Mauro Bendinelli, *University of Pisa*
Herman Friedman, *University of South Florida College of Medicine*

Recent volumes in the series:

DNA TUMOR VIRUSES
Oncogenic Mechanisms
Edited by Giuseppe Barbanti-Brodano, Mauro Bendinelli, and Herman Friedman

ENTERIC INFECTIONS AND IMMUNITY
Edited by Lois J. Paradise, Mauro Bendinelli, and Herman Friedman

FUNGAL INFECTIONS AND IMMUNE RESPONSES
Edited by Juneann W. Murphy, Herman Friedman, and Mauro Bendinelli

HERPESVIRUSES AND IMMUNITY
Edited by Peter G. Medveczky, Herman Friedman, and Mauro Bendinelli

MICROORGANISMS AND AUTOIMMUNE DISEASES
Edited by Herman Friedman, Noel R. Rose, and Mauro Bendinelli

NEUROPATHOGENIC VIRUSES AND IMMUNITY
Edited by Steven Specter, Mauro Bendinelli, and Herman Friedman

PSEUDOMONAS AERUGINOSA AS AN OPPORTUNISTIC PATHOGEN
Edited by Mario Campa, Mauro Bendinelli, and Herman Friedman

PULMONARY INFECTIONS AND IMMUNITY
Edited by Herman Chmel, Mauro Bendinelli, and Herman Friedman

RAPID DETECTION OF INFECTIOUS AGENTS
Edited by Steven Specter, Mauro Bendinelli, and Herman Friedman

RICKETTSIAL INFECTION AND IMMUNITY
Edited by Burt Anderson, Herman Friedman, and Mauro Bendinelli

VIRUS-INDUCED IMMUNOSUPPRESSION
Edited by Steven Specter, Mauro Bendinelli, and Herman Friedman

---

Rapid Detection of Infectious Agents

Edited by

Steven Specter
University of South Florida College of Medicine
Tampa, Florida

Mauro Bendinelli
University of Pisa
Pisa, Italy

and

Herman Friedman
University of South Florida College of Medicine
Tampa, Florida

Plenum Press • New York and London

Library of Congress Cataloging-in-Publication Data

On file

ISBN 0-306-45848-9

A Division of Plenum Publishing Corporation
233 Spring Street, New York, N.Y. 10013

http://www.plenum.com

10 9 8 7 6 5 4 3 2 1

Printed in the United States of America

Contributors

BURT ANDERSON • College of Medicine, Department of Medical Microbiology and Immunology, University of South Florida, Tampa, Florida 33612-4799

M. N. BOBROW • Dupont-NEN Life Science Products, Medical Products Department, Boston, Massachusetts 02118

CHARLOTTE A. GAYDOS • Division of Infectious Disease, The Johns Hopkins University School of Medicine, Baltimore, Maryland 21205

RICHARD V. GOERING • Department of Medical Microbiology and Immunology, Creighton University School of Medicine, Omaha, Nebraska 68178

SPENCER R. HEDGES • Department of Microbiology, University of Alabama at Birmingham, Birmingham, Alabama 35294-2170

RICHARD L. HODINKA • Departments of Pathology and Pediatrics, Clinical Virology Laboratory, Children's Hospital of Philadelphia, and University of Pennsylvania School of Medicine, Philadelphia, Pennsylvania 19104

SUSAN B. HUNTER • Foodborne and Diarrheal Diseases Branch, Division of Bacterial and Mycotic Diseases, National Center for Infectious Diseases, Centers for Disease Control and Prevention, Atlanta, Georgia 30333

G. J. LITT • Dupont-NEN Life Science Products, Medical Products Department, Boston, Massachusetts 02118

JAMES J. McSHARRY • Department of Microbiology, Immunology, and Molecular Genetics, Albany Medical College, Albany, New York 12208

JIRI MESTECKY • Department of Microbiology, University of Alabama at Birmingham, Birmingham, Alabama 35294-2170

SUSANNE MODROW • Institute for Medical Microbiology and Hygiene, University of Regensburg D-93053 Regensburg, Germany

GERARD J. NUOVO • MGN Medical Research Laboratory, Setauket, New York 11733

PAUL D. OLIVO • Department of Molecular Microbiology, Washington University School of Medicine, St. Louis, Missouri 63110

DAVID PERSING • Division of Clinical Microbiology, Mayo Clinic Foundation, Rochester, Minnesota 55905

THOMAS C. QUINN • Divison of Infectious Disease, The Johns Hopkins University School of Medicine, Baltimore, Maryland 21205; and National Institute of Allergy and Infectious Diseases, National Institutes of Health, Bethesda, Maryland 20892-2520

MICHAEL W. RUSSELL • Department of Microbiology, University of Alabama at Birmingham, Birmingham, Alabama 35294-2170

BALA SWAMINATHAN • Foodborne and Diarrheal Diseases Branch, Division of Bacterial and Mycotic Diseases, National Center for Infectious Diseases, Centers for Disease Control and Prevention, Atlanta, Georgia 30333

DANNY L. WIEDBRAUK • Departments of Clinical Pathology and Pediatrics, William Beaumont Hospital, Royal Oak, Michigan 48073

HANS WOLF • Institute for Medical Microbiology and Hygiene, University of Regensburg, D-93053 Regensburg, Germany

XIAOTIAN ZHENG • Division of Clinical Microbiology, Mayo Clinic Foundation, Rochester, Minnesota 55905

Preface to the Series

The mechanisms of disease production by infectious agents are presently the focus of an unprecedented flowering of studies. The field has undoubtedly received impetus from the considerable advances recently made in the understanding of the structure, biochemistry, and biology of viruses, bacteria, fungi, and other parasites. Another contributing factor is our improved knowledge of immune responses and other adaptive or constitutive mechanisms by which hosts react to infection. Furthermore, recombinant DNA technology, monoclonal antibodies, and other newer methodologies have provided the technical tools for examining questions previously considered too complex to be successfully tackled. The most important incentive of all is probably the regenerated idea that infection might be the initiating event in many clinical entities presently classified as idiopathic or of uncertain origin.

Infectious pathogenesis research holds great promise. As more information is uncovered, it is becoming increasingly apparent that our present knowledge of the pathogenic potential of infectious agents is often limited to the most noticeable effects, which sometimes represent only the tip of the iceberg. For example, it is now well appreciated that pathologic processes caused by infectious agents may emerge clinically after an incubation of decades and may result from genetic, immunologic, and other indirect routes more than from the infecting agent in itself. Thus, there is a general expectation that continued investigation will lead to the isolation of new agents of infection, the identification of hitherto unsuspected etiologic correlations, and, eventually, more effective approaches to prevention and therapy.

Studies on the mechanisms of disease caused by infectious agents demand a breadth of understanding across many specialized areas, as well as much cooperation between clinicians and experimentalists. The series *Infectious Agents and Pathogenesis* is intended not only to document the state of the art in this fascinating and challenging field but also to help lay bridges among diverse areas and people.

M. Bendinelli
H. Friedman

Preface

1. INTRODUCTION

The evolution of diagnostic microbiology began with the development of the microscope and has progressed through a variety of tools that have allowed us initially to visualize microorganisms and then to take advantage of various properties of the bacteria, viruses, fungi, or parasites to aid in their diagnosis. Earliest techniques relied on describing gross or microscopic morphology and progressed to observing growth on differential and selective media. Both of these methods depended on visualizing the organisms and provided fairly rapid detection, usually within 24 hours. More powerful tools, like the electron microscope, allowed visual identification of microbes including viruses within 8–24 hours. Additionally, the development of cell culture techniques facilitated detecting viruses and other obligate intracellular parasites, yet detection required from 1 day to a few weeks in some cases. However, although effective for gross identification, these tools still were often too crude to distinguish microorganisms with like morphologies.

Subsequently, immunologic and biochemical tools that differentiated species and strains were developed, and diagnostic microbiology became a highly specific science, sometimes providing diagnosis within one or two days. Furthermore, automation allowed for diagnosis in some cases in less than 24 hours. Serological profiles that depend on detecting multiple antigens and/or antibodies or the presence of IgM antibodies, as an indication of primary infection, also aided detection and identification within hours of specimen submission. More sophisticated immunologic tools also are being applied in the clinical lab, such as new uses for flow cytometry, development of synthetic peptides as reagents for diagnostic testing, and measuring antibodies in mucosal secretions.

The development of molecular biology has expanded exponentially in the past decade and has begun to be accorded rapidly increasing importance in the diagnostic laboratory. This has tremendously improved our ability to recognize even small differences among isolates, has greatly refined our diagnostic skills, and has truly made almost all aspects of diagnostic microbiology rapid.

Many of the methods developed for detecting microorganisms depend on the viability of the organisms. Many conventional serological techniques, although not suffering this disadvantage, are often limited by the need for matched, timely collected, acute and convalescent sera, which delays diagnosis, often by weeks to months. Thus, there is a tremendous advantage in using direct detection methods for rapid diagnosis of infectious diseases. This has been achieved by a variety of methods and tools. Our increasing sophistication has allowed detecting small amounts of infectious material, achieving even more rapid identification and often providing positive identification in a few hours. Additionally, we have expanded our capabilities now to distinguish strains within a species and specific isolates. This has resulted in great improvements in epidemiological studies. Our expanding repertoire of techniques has also increased the diversity of information that we can obtain and therefore the contribution of the diagnostic laboratory to guide patient management effectively. Now techniques allow quantitation of viral loads, which has been linked to pathogenic activity for the human immunodeficiency virus (HIV), hepatitis C virus (HCV), and cytomegalovirus (CMV), among others, chronically present once infection is established. Genotyping has been most important for relating to therapeutic protocols, especially for HCV. Antiviral susceptibility testing has become a reality with the advent of new techniques that rapidly detect and quantitate virus and also probe for resistance mutations in the viral genome.

2. IMMUNOLOGIC ADVANCES

Serological and immunologic assays have been utilized extensively in diagnosing infectious diseases, in many cases for rapid diagnosis. Yet many of these assays depend on paired samples, which significantly delays testing. Furthermore, serological tests often depend on developing measurable levels of antibody, which takes anywhere from 10 days to several months, depending on the pathogen and the individual patient response. The development and use of monoclonal antibodies has greatly expanded our ability to perform direct antigen detection testing, one of the singularly most important steps in developing rapid diagnostic services. Parallel to this is the more recent development of synthetic peptides that can be used to dissect and more specifically measure immune responses. These allow rapid detection and greater definition of the antigenic epitopes responsible for inducing immunity. Although synthetic peptides are also likely to have a great impact on vaccine development and understanding of immunodominant epitopes, there are additional uses for these peptides in identifying epitopes responsible for generating immunity during infections. Recent developments and use of synthetic peptides in infectious disease diagnosis are described by Susanne Modrow and Hans Wolf of the Institute for Medical Microbiology and Hygiene of Regensburg, Germany.

A most interesting development in the use of immunologic methods for rapidly diagnosing viral infections is the adaptation of flow cytometry for such use. As described by Jim McSharry, Albany Medical College, Albany, New York, this method is reproducible, rapid, and has a variety of applications that can help in diagnosing several viral diseases. Furthermore, it has been utilized to assist in determining drug efficacy in HIV infections.

Detecting antibodies in fluids other than serum has become a more important diagnostic approach in recent years. In this regard measuring mucosal antibodies is gaining in significance, most notably secretory IgA. The manner of detection and utility of this form of testing is detailed in the chapter by Jiri Mestecky and co-workers, University of Alabama at Birmingham, Birmingham, Alabama.

3. CELL CULTURE AND GENETIC ENGINEERING

Cell culture has been the gold standard for viral isolation for the past 50 years. More recently the development of the shell vial and enhanced cell culture using labeled antibodies for detecting virus has brought this into the realm of rapid detection. The newest advance in cell culture is the use of genetically engineered cell lines that express a gene, such as the beat-galactosidase gene, when infected only by a particular virus. This results in a colorimetric reaction (cells turn blue when infected) that self-identifies an infected cell without adding antibodies. This mode of rapid detection is described by Paul Olivo, Washington University, St. Louis, Missouri, who adapted this method for detecting herpes simplex virus in clinical specimens. This approach is being adapted for other viruses and it might be possible in the future to have engineered cells that have several such reporter genes each of which expresses a different color in response to a particular virus. Thus, by expressing cell color, we might isolate and identify viruses rapidly in one cell type in a very futuristic technique. Dr. Olivo also describes alternative uses for genetic engineering of cells, such as transgenic cell lines or cell lines designed to resist one virus but not another, to aid in differentiating viruses in mixed infections or other uses.

4. MOLECULAR BIOLOGY AND RAPID DETECTION

The remainder of the work presented in this volume utilizes molecular techniques which are applied for rapidly detecting and characterizing a variety of organisms. Xiaotian Zheng and David Persing, Mayo Clinic, Rochester, Minnesota, examine a variety of these techniques comparing the different strategies that are used for genetic amplification to make detection possible. Techniques discussed encompass target amplification, probe amplification, signal amplification, and amplifications that utilize quantitative measurements. Clearly, the most widely used of these is the polymerase chain reaction (PCR). The application of this technique has reached virtually every aspect of microbiology in the research lab and is now being applied to clinical specimens on an ever expanding basis, as described by Danny Wiedbrauk, William Beaumont Hospital, Royal Oak, Michigan, and Richard Hodinka, Children's Hospital of Philadelphia. In addition, a variety of adaptations of PCR have been utilized in more demanding settings.

Gerard Nuovo, MGN Medical Research Laboratory, Setauket, New York, details how PCR can be used *in situ* to detect viral infections in tissues. This has proven useful for detecting HIV-1 in lymph nodes, brain cells, and spermatogonia, assisting in helping to locate latent virus or indicate potential for sexually transmitting virus. Broad range PCR is described by Burt Anderson, University of South Florida College of Medicine, Tampa, Florida. This procedure is useful for examining large genomes and a description is presented on how this is applied to identifying novel bacteria not readily cultivated.

Similar to broad range PCR, Richard Goering, Creighton University School of Medicine, Omaha, Nebraska, describes molecular epidemiology for nosocomial infections using methods that are effective in identifying the large genomes of bacteria. The primary method described in this chapter is pulsed-field gel electrophoresis (PFGE), which has significant advantage over other methods for identifying responsible agents in epidemics in hospital.

Additionally, two other methods of amplification are described in this volume. Charlotte Gaydos and Thomas Quinn, The Johns Hopkins University School of Medicine, Baltimore, Maryland present the ligase chain reaction for detecting sexually transmitted diseases (STD). This enzymatic method for nucleic acid amplification is an effective alternative to PCR and is being commercially developed to compete in molecular diagnosis with the better known PCR. Its application for STDs makes the genitourinary specimen popular for using this methodology. Finally, the desire to develop nonradiometric methods of signal amplification has led to the development of tyramide signal amplification, as described by Gerald Litt and M. Bobrow, Dupont-NEN Life Science Products, Boston, Massachusetts. Rather than amplifying the microorganism's nucleic acid, this methodology amplifies the signal substrate in an enzyme-based assay. In this regard tyramine is coupled to a substrate for horseradish peroxidase. Thus the signal in an immunoassay is amplified thereby increasing assay sensitivity. The advantages of the method include the use of small amounts of reagents, such as scarce primary antibodies, leading to saving of resources, costs, and time to perform studies.

5. PERSPECTIVE

The goal of rapid diagnosis for all infectious disease has yet to be achieved. However, through the use of newer immunologic and molecular biological tools we have made great advances in the past decade toward this goal. The adaptation of research laboratory methods to routine laboratory diagnosis of infectious diseases has progressed at a rate far faster than one might have anticipated at the start of the 90s. Diagnosing a number of fastidious organisms is now possible using molecular diagnostics, quantitation of organisms where numbers are related to clinical disease is now achievable, monitoring of antiviral susceptibilities has been enhanced, and epidemiological studies have been greatly improved. This has, however, created new concerns for the laboratory microbiologist. Increased sensitivity of detection methods has generated the problem of interpreting the clinical significance of the presence of microbial nucleic acid or the antigen of an infectious agent in an individual. This is especially true for persistent and ubiquitous agents, such as herpesviruses, adenoviruses, HIV, etc., which are detected in the absence of active disease or during acute diseases that have another etiology. In these cases quantitating the infectious burden is helpful, but the need often exists for additional assays and new criteria to define the clinical significance of the organism detected in the patient under investigation. Thus, future efforts should focus on obtaining data that help define this clinical significance and on rendering current assays simpler and less expensive to perform.

The pages that follow demonstrate how we have achieved many successes in the rapidly diagnosing infection and providing a foundation for taking these and techniques yet to be developed to a level where diagnosis for all infections are achieved rapidly and effectively.

S. Specter
M. Bendinelli
H. Friedman

Acknowledgments

The editors of this volume thank all contributors for each of the reviews for their time and effort in preparing such excellent chapters. The editors also thank Ms. Ilona M. Friedman for invaluable assistance as editorial coordinator and managing editor for this series. We also thank the editorial and production staff of Plenum Publishing Corporation for their outstanding assistance. It is anticipated that publication of this volume will focus attention on the rapid advances in the area of diagnosis of viral infections.

Contents

7. Applications of the Polymerase Chain Reaction

DANNY L. WIEDBRAUK AND RICHARD L. HODINKA

8. Identifying Novel Bacteria Using a Broad-Range Polymerase Chain Reaction

BURT ANDERSON

1

Genetically Engineered Cells in Diagnostic Virology

PAUL D. OLIVO

1. INTRODUCTION

Rapid diagnostic assays based on direct detection of viral antigen or nucleic acid are being used with increasing frequency in clinical virology laboratories. Viral culture is the only way to detect infectious virus and to analyze clinically relevant viral phenotypes, such as drug resistance, but it is labor-intensive, time-consuming, and requires the use of many different cell lines.[1] Transgenic technology, together with increasing knowledge of the molecular pathways of viral replication, offers the possibility of using genetically modified cell lines to improve viral growth in cell culture and to facilitate rapid detection of virally-infected cells. Genetically modifying cells so that they express an easily measured reporter enzyme only after infection with a specific virus allows the detection of infectious virus by using rapid and simple enzyme assays, such as β-galactosidase, without the need for antibodies. Although transgenic cells have recently been successfully used for herpes simplex virus (HSV) detection, much more work needs to be done to adapt this technology to other human viral pathogens, such as cytomegalovirus (HCMV) and respiratory viruses. This review offers some strategies for applying this technology to a wide spectrum of animal viruses.

PAUL D. OLIVO • Department of Molecular Microbiology, Washington University School of Medicine, St. Louis, Missouri 63110.

Rapid Detection of Infectious Agents, edited by Specter *et al.* Plenum Press, New York, 1998.

1.1. History of Viral Culture

The development of animal cell culture techniques was critical to the field of animal virology because it provided a convenient method for growing and quantitating animal viruses. The initial reports on the growth of poliovirus in cell culture and development of the plaque assay paved the way for virtually all of the major advances in animal virology in the past 40 to 50 years.[2,3] In addition, the field of medical virology and the importance of the clinical virology laboratory in large part grew out of progress in identifying human viral pathogens through the recognition of specific virally-introduced cytopathic effects (CPE) in cultured cells. A large body of information about the suitability of particular cell lines for detecting human pathogenic viruses has been accumulated, but there is an ongoing search for more sensitive cell culture systems for detecting pathogenic viruses including those currently not cultivatable.[4–6]

1.2. Viral Replication, Host Range, Susceptibility, and Permissivity

Viruses are obligate intracellular pathogens and replication of viruses in cells results from complex interactions between virion-associated factors, virally-encoded factors, and host-cell factors. Advances in recombinant DNA technology have expanded our knowledge of molecular mechanisms of viral replication, including knowledge of viruses that cannot be propagated in cell culture. It was clear from early attempts to propagate viruses that not all cell types would support the replication of a given virus and that this is related to the host range and tissue tropism of the virus. The species from which the cell is derived, the lineage of the cell, and the degree of differentiation, for example, have profound effects on whether or not a particular cell line supports replication of a particular virus.[7]

The replicative cycle of most viruses can be described in terms of distinct stages, each of which requires specific molecular events that involve specific host-cell factors, which interact with viral elements in precise ways. As knowledge of many viral life cycles has increased, the specific cellular factors that are critical for viral replication are being identified.[8–16]

There are many reasons why a virus might not replicate in a certain cell type, but broad categories account for the majority of cases: (1) inability of the virus to enter the cell and (2) inability of the virus to express its genes and to replicate its genome once it enters the cell. Inability of the virus to enter a cell results from many factors, but the absence of the appropriate virus receptor is probably the most common reason.[17] Once a virus binds to the cell surface, complex interactions between viral proteins and cell surface molecules must occur to allow the virus to enter the cell by either fusion or endocytosis.[8,18]

After a virus enters the cell, it relies on many cellular functions to express its genes, replicate its genome, and package replicated genomes into new virions. Many of these cellular functions are general and ubiquitous. Energy-producing organelles and protein synthesis machinery, for example, provide functions that all obligate intracellular microorganisms require. However, an increasing number of reports find that specific cellular factors or enzymes, not found in all cells, are used by particular viruses and that the presence or absence of these factors dictate whether or not a virus undergoes a complete replicative cycle. For example, cellular protein kinases play a vital role in gene expression of rhabdoviruses, such as vesicular stomatitis virus (VSV), and paramyxoviruses, such as respiratory syncytial virus (RSV) and parainfluenza viruses. A cascade of P protein phosphorylation is apparently important in forming an active RNA polymerase complex. In VSV and RSV, the ubiquitous casein kinase II is utilized, but parainfluenza virus type 3 requires the more selectively expressed cellular protein kinase C isoform ζ.[16] For DNA viruses and retroviruses, nuclear transcription factors are selectively utilized to express vital genes and in many cases these factors are not expressed in all cells.[7,19] The inability of polyomavirus to replicate in undifferentiated teratocarcinoma cells was traced to low amounts of the transcription factor AP1 that acts on the viral enhancer sequence.[20] As another example, the tissue specificity of hepatitis B virus (HBV) may partially result from the activity of the hepatocyte-specific transcription factor, hepatocycte nuclear factor 3 (HNF3), which acts on the viral enhancer.[21] Specific components of DNA replicative enzymes account for species-specific permissivity as is the case for mouse polyomavirus which requires the mouse p48 subunit of the DNA polymerase alpha-primase complex for replicating its DNA genome.[14]

2. STABLE TRANSFORMATION OF CELLS TO CONFER VIRAL SUSCEPTIBILITY

The stable introduction of foreign genetic material into a cell has become a routine paradigm in molecular biology laboratories today, and it is an important tool for viral geneticists.[22] This methodology combined with the knowledge of specific virus–cell interactions is used to design and engineer cells to become susceptible, permissive, or even resistant to particular viruses or viral families. This possibility frees virologists from relying on fortune and serendipity to identify the best cell lines for viral studies. One could view this approach as analogous to the strategy of drug design in place of random screening to identify effective pharmaceutical agents.

A number of reports have demonstrated that cells lacking a viral receptor are transformed by the gene for this receptor and, in doing so, acquire susceptibility to the virus. This transformation has been used as one of the strategies to identify a number of viral receptors. The gene for the poliovirus receptor (PVR) was isolated by transforming nonsusceptible mouse L cells with a cDNA library from susceptible HeLa cells.[23] Then clones of L cells susceptible to poliovirus were used to isolate the gene for the poliovirus receptor which set the stage for identifying the receptor itself. This technology has been applied to viral detection. A recent report describes the characterization of mouse L cells (L20B) expressing the human poliovirus receptor for the specific detection of polioviruses *in vitro*.[24] These authors noted that the transgenic L20B cells are only marginally less sensitive for detecting polioviruses compared with human HEp2C cells but have the advantage of being nonpermissive for other enteric human picornaviruses.

Cell lines have been genetically modified to stably express the CD4 molecule which is the high affinity human immunodeficiency virus (HIV) receptor.[25,26] Mouse cell lines and certain human cells that have been modified to express CD4, however, remain unsusceptible to HIV, because apparently they lack a cellular factor, found on many human cells, that is required for viral entry.[18] CD4-expressing HeLa cells are susceptible to HIV and have been used in developing a sensitive, quantitative, focal assay for HIV infectivity.[27] Unfortunately, primary isolates of HIV-1 do not infect CD4-expressing HeLa very efficiently because they lack a cell-surface fusion cofactor, such as CCR5, required for entry of macrophage-tropic isolates.[28–30]

Among the herpesviruses, Epstein–Barr virus (EBV) is the only virus for which a clearly defined high-affinity receptor, the CR2 molecule, has been identified.[31] CR2 is a glycoprotein found in the plasma membrane of B lymphocytes and certain epithelial cells. It is a complement receptor that binds the C3d component of complement and is thought to be involved in lymphocyte activation. The presence of CR2 on the cell surface plays a major role in EBV infection, but other host factors, found only in certain primate cells, are important in determining susceptibility of cells to EBV and in determining whether an infection is productive, abortive, or latent.[32] There is no reliable and efficient cell culture system for propagating EBV. Infection of B cells *in vitro* causes a predominantly latent infection and immortalization of cells. The only method presently available for demonstrating infectious virus in a patient sample is the lymphocyte immortalization test.[33] Unfortunately, it is cumbersome to perform and, therefore, not routinely available. Although *in vitro* infection of primary epithelial cell cultures leads to a productive infection, it is very inefficient in part because only a small fraction of primary human epithelial cells express CR2. Recently, it was reported that epithelial cell lines lacking the CR2 receptor could be made permissive for EBV by transforming them to stably express CR2 on their surfaces.[34] The efficiency of the infection is not high, but this work represents an encouraging step in the developing cell lines useful for propagating and detecting EBV.

3. TRANSGENIC CELLS THAT FACILITATE VIRAL DETECTION

3.1. General Principles

A number of the molecular processes that are involved in viral replication are unique to virally-infected cells. Transcription from certain viral promoters and replication using viral enzymes are examples of events that do not occur in uninfected cells but are critical for viral replication. Such virus-specific events are potential targets for novel antiviral agents, but they are also used to identify infected cells. The general notion is to take advantage of the fact that all viruses have unique pathways for expressing their genes, and by using our knowledge of the specific signals of these pathways, we can induce the expression of a reporter gene. The basic strategy is to stably introduce genetic elements into a cell such that when a virus, and only a particular virus, enters this cell, a virus-specific event is triggered that results in producing an easily measurable enzyme (Fig. 1). Thus, this strategy provides a simple and virus-specific detection system. The genetic elements are derived from viral, bacterial, and cellular sources. The acronym applied to the overall paradigm is Enzyme-Linked Virus Inducible System or ELVIS™. Because of fundamental differences in their molecular biology, DNA and RNA viruses, require different strategies to promote enzyme induction. These strategies are outlined later.

As originally pointed out by Baltimore, all viral replication involves efficient pathways for generating mRNA to be translated into viral proteins.[35] In fact, viruses are classified according to how they derive their early mRNA and, in turn, this, is largely dictated by the type of nucleic acid in the viral genome. By taking advantage of knowledge of the signals involved in these pathways, mRNA is generated for a reporter gene in virally-infected cells. Theoretically, this can be accomplished for all viral families. In addition, many viruses have specific and unique pathways for promoting translation of their mRNAs and for proteolytically processing viral polyproteins into functional proteins. It is likely that these pathways also can be subverted to convert inactive reporter enzymes into functional enzymes in virally-infected cells. This approach to viral detection is now feasible for viruses from several different families including a retrovirus (HIV), a DNA virus (HSV), and an RNA virus (Sindbis virus).

3.1.1. *Viruses and Viruses that Replicate through a DNA Intermediate (e.g., Retroviruses)*

As mentioned previously, DNA viruses and RNA viruses have fundamental differences in their replicative pathways. It is helpful to discuss these viral groups separately when describing the principals involved in generating viral detection

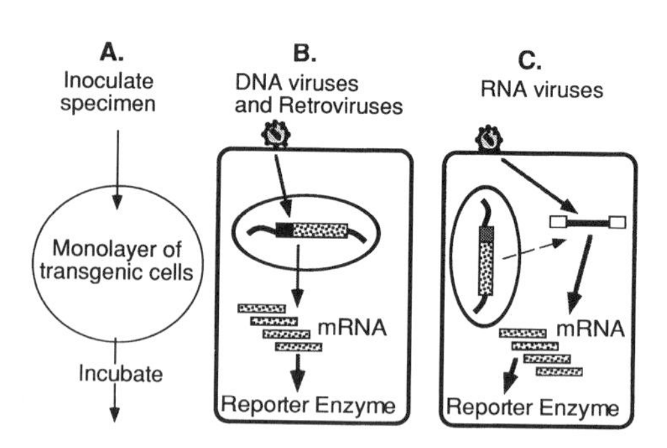

FIGURE 1. Schematic representation of the concept of detecting virally infected cells using a culture of transgenic cells in which a reporter gene is expressed only upon induction by a specific virus. (A) Specimens are handled similarly to a viral culture, but the detection method involves measuring a reporter enzyme rather than observing for cytopathic effects or using immunological reagents. (B) A strategy useful for detecting DNA viruses and retroviruses both of which replicate in the nucleus. Cells are stably transformed with one or more copies of a chimeric gene consisting of a reporter gene (▩) downstream of a viral promoter (■). Ideally, in uninfected cells the promoter is inactive, no reporter gene mRNA is transcribed, and thus no reporter enzyme is produced. After infection with the virus from which the promoter is derived, virus transactivator proteins act with cellular RNA polymerase to transcribe many copies of reporter gene mRNA which are transported to the cytoplasm where they are translated into high levels of an easily measured reporter enzyme. (C) A strategy useful for detecting RNA viruses. Cells are stably transformed with one or more copies of a DNA segment consisting of a reporter gene (▩) flanked by cDNA from the 5′ and 3′ regions of an RNA viral genome (□) which contain the cis-acting elements required for viral RNA replication and transcription. Upstream of the 5′ viral cDNA is a cellular promoter (▦) that is active in uninfected cells which constitutively express an RNA molecule, which for structural reasons cannot serve as mRNA for the reporter gene even though it contains its open reading frame. After infection with the virus from which the cDNA was derived, the viral RNA polymerase and other nonstructural proteins replicate and transcribe the constitutively expressed chimeric RNA molecule thus generating high levels of reporter gene mRNA which is translated into high levels of an easily measured reporter enzyme.

cell lines. DNA viruses (except for cytoplasmic DNA viruses, such as poxviruses) and viruses that go through a DNA intermediate during replication (i.e., retroviruses) derive their mRNA from transcription carried out in the nucleus using cellular DNA-dependent RNA polymerase in conjunction with cellular and viral transcription factors. The specificity of viral transcription depends largely on DNA sequence elements upstream of viral genes (e.g. promoter and enhancer elements). A strategy commonly used by molecular biologists to study viral promoters is to generate chimeric genes in which a so-called reporter gene is placed downstream of, and thus under the regulatory control of, a viral promoter, often with additional 3′ sequences from a eukaryotic gene to promote mRNA processing (e.g., polyadenylation). Then such chimeric DNA constructs are used to study the activity of the promoter either in transient transfective assays or by placing the chimeric gene in the viral genome. Recently, several reports have described isolating cell lines that have been stably transformed with such chimeric genes using herpesvirus or retrovirus promoters (see later).[36–40] Viral transactivator proteins, either brought into the cell with the virion or made *de novo* in infected cells, act in concert with cellular RNA polymerase and transcription factors to upregulate expression of the chimeric gene which is manifested by an increase in reporter enzyme activity (Fig. 1). These reports note that such cell lines are convenient indicators of the viral presence in an inoculum by virtue of the ease with which the reporter enzyme is measured.

The key question with this type of viral detection system is, How useful is it for a virology laboratory and particularly a clinical virology laboratory? As with any detection assay, the critical issues are sensitivity and specificity. These are in turn affected by the three primary variables of the system: (1) the viral promoter, (2) the cell type, and (3) the reporter gene. The behavior of the viral promoter is centrally important to both the sensitivity and specificity of this type of viral detection system. Three key questions are (1) Is the promoter "quiet" in uninfected cells, i.e., is there constitutive expression from this promoter? (2) To what degree is the promoter upregulated by the viral transactivator proteins? and (3) Can heterologous viral transactivator proteins activate the promoter? The level of constitutive expression and the degree of transactivation of the viral promoter is primarily a function of the intrinsic properties of the promoter, and thus knowledge of the properties of a given viral promoter influences the choice of promoter. However, the behavior of viral promoters is clearly influenced by the cell type, the location of the chimeric gene in the cellular genome, and the number of chimeric gene copies. In fact, it is quite apparent that the behavior of viral promoters within a viral genome does not always predict how they will behave outside the viral genome, whether in transient assays or when stably integrated into a host chromosome. Isolating a cell line with properties well suited to viral detection, therefore, involves a large measure of empiricism.

The cell type is a critical variable. The susceptibility and permissibility of the cell line for various viruses dramatically affect the sensitivity and the specificity of such a viral detection assay. For the assay to work, it is necessary that the virus being detected bind to the cell, enter the cell, and initiate its replicative cycle (i.e., the cell must be susceptible to the virus). The degree of susceptibility of a cell line for a particular virus has a profound effect on the assay's sensitivity. However, it is not necessarily important that the virus undergo a full round of replication in the cell (i.e., the cell need not be permissive) because triggering the reporter gene via the viral promoter can be an early event during viral replication. In fact, this may broaden the number of cell lines available for detecting a given virus because, unlike a CPE-based assay, susceptible, but nonpermissive cell lines can be used. On the other hand, the lack of susceptibility of a cell line for heterologous viruses (viruses other than the virus from which the promoter is obtained), contributes to the specificity of the system. Thus specificity need not completely rely on the lack of transactivation of the promoter by heterologous viruses and can be conferred by the cell type.

Finally, the reporter gene chosen can have significant effects on the sensitivity, specificity, and versatility of the system. A number of so-called reporter genes have been used by molecular biologists and the convenience of measuring the enzymatic activities of their gene products is constantly increasing as more substrates are isolated and assays are developed.[41] In fact, one of the attractive features of this technology is that there are a large number of permutations possible, given the choices of reporter enzymes, substrates, and types of assays (Fig. 1). Although antibody-based methods utilize some of these same enzymes, such methods involve significantly more sample manipulation and are not as versatile. *E coli* β-galactosidase is a commonly used reporter enzyme which has the advantage of being measurable in several different formats using various commercially available substrates. However, endogenous β-galactosidase activity is found to varying degrees in certain cell types or is found in bacteria which might contaminate a specimen. Firefly luciferase has the double advantage of being measurable by one of the most sensitive assays available and of being absent in virtually all mammalian cell lines.[42] This allows the possibility of a signal-to-noise ratio sufficiently high that a single infectious virus is detectable (see later).[43,44] Green fluorescent protein (GFP), especially the recently improved derivatives, offer the possibility of detecting virally-infected cells with a minimum of manipulation because GFP is inherently fluorescent and requires no substrates or cofactors.[45]

3.1.2. RNA Viruses

Replication of RNA is a process unique to RNA viruses that replicate their RNA genomes and generate their mRNA using a virally encoded RNA-dependent RNA polymerase and other nonstructural viral proteins. Therefore, the strategy described previously of using DNA promoters which are acted upon by cellular DNA-dependent RNA polymerase does not work with RNA viruses. A different strategy for triggering expression of a reporter gene in RNA virus-infected cells must be used. For those RNA viruses that have been intensively studied, it has been shown that viral RNA-dependent polymerase complexes use specific nucleotide sequences found on the viral RNA to direct RNA replication and transcription. Therefore, one can genetically modify a cell so that it constitutively expresses low levels of an RNA molecule which, although it contains the open reading frame (ORF) of a reporter gene, is not translated into a reporter enzyme because its ORF is far downstream of the 5′ end of the RNA molecule. This RNA molecule, however, can be strategically constructed so that it contains two viral sequence elements: (1) sequence elements at both the 5′ and 3′ ends of the molecule that are essential for replicating the RNA by viral RNA-dependent RNA polymerase and (2) internal sequence elements that are placed immediately upstream of the reporter gene ORF and which promote RNA to RNA transcription by the viral RNA polymerase to generate reporter gene mRNA. Cells infected with the virus from which the sequence elements are derived, however, would exhibit high levels of reporter enzyme because infection would initiate high replicative levels of the RNA molecule and efficient transcription of many copies of reporter gene mRNA which then would be translated into reporter enzyme [Fig. 1(C)].

One potential problem with this strategy is that our knowledge base does not allow us to predict confidently whether the genetically modified cells would properly transcribe the designed RNA molecule so that it would have the correct 5′ and 3′ ends or whether the RNA would be transported intact into the cytoplasm of the cell so as to encounter the viral RNA polymerase and be replicated in infected cells. As with the strategy outlined previously for DNA viruses, therefore, generating cell lines which are useful in RNA detection assays will involve a significant amount of empiricism. In addition, the issue of specificity and sensitivity which were discussed in the context of DNA viral detection assays, particularly as they relate to the viral nucleotide sequence elements, the cell type, and the reporter gene, are fully applicable to cell-line-based assays for RNA viruses. A prototypical cell line that demonstrates the validity of this design was recently reported and is described in detail later.[43]

3.2. Application to Specific Viral Families

3.2.1. *Retroviruses*

3.2.1a. Human immunodeficiency virus. A quantitative bioassay for HIV-1 was developed from a CD4-positive lymphoid cell line transformed with a retroviral vector that contains an HIV-1 LTR promoter linked to the chloramphenicol acetyltransferase (CAT) gene.[36] The HIV LTR promoter contains sequences that provide the promoter with the ability to be transactivated by the HIV *tat* gene product. It was reported that when such cell lines are infected with HIV-1 or co-cultivated with HIV-1 infected cells, CAT activity increases 500-1000-fold. Ten fold lower levels of CAT activity were measured after infection with simian immunodeficiency viruses, suggesting that these viruses have similar, but distinct, transactivating proteins. HTLV-1, HTLV-2, equine infectious anemia virus, and HSV-1 do not activate CAT above the background levels of uninfected cells. The authors also demonstrated that their indicator cell lines could be used to quantitate HIV because when the indicator cell line are in excess, a linear relationship exists between the amount of infectious virus and the level of CAT activity within at least two orders of magnitude. Finally, these investigators showed the potential of using such indicator cell lines in rapid *in vitro* screening assays for evaluating anti-HIV agents.

Several laboratories have subsequently reported using a histochemical β-galactosidase assay for detecting and analyzing HIV-infected cells.[37–39] The strategy was based on a HeLa cell line that is rendered susceptible to HIV transformation with the CD4 gene.[27] Then CD4-expressing HeLa cells (HT4 cells) were stably transformed with a DNA fragment containing a recombinant HIV-1, HIV-2, and SIV-infected cells, but not HTLV-1. In these systems, problems with uninfected-cell β-galactosidase activity were circumvented by adding a nuclear localization signal to the β-galactosidase. Nevertheless, a single virion or a single infected lymphoid cell leads to a cluster of infected cells or a syncytium which histochemically stains positive for β-galactosidase. However, a fluorometric β-galactosidase assay on cell lysates was relatively insensitive. Specificity for HIV was not reported, but would likely be a problem because other viruses, such as HSV, transactivate the HIV LTR promoter.[46–48] In another report it is suggested that this type of HIV-detection cell line is very useful in antiretroviral drug screening assays.[49] As mentioned before, a significant limitation of HT4 cell-based detection systems is that HT4 cells are poorly infected by primary HIV-1 isolates because they lack certain cell-surface cofactors.[28–30]

3.2.2. Herpesviruses

3.2.2a. Herpes Simplex Virus. HSV types 1 and 2 (collectively HSV) infect a large number of individuals each year, and the diagnosis of HSV infection is one of the more frequent tests performed by diagnostic virology laboratories. Furthermore, effective therapy for HSV has stimulated the need for more sensitive, accurate, and rapid diagnostic test. Recently, a cell line (BHKICP6LacZ) was described which is stably transformed by the *E. coli lacZ* gene behind the HSV-1 *UL39* promoter.[40] After infection, HSV-infected cells are detected by a number of different β-galactosidase assays, including a simple colorimetric assay of cell lysates. Microscopic assays using fluorogenic substrates and fluorescent microscopy or a simple histochemical stain and light microscopy detect individual infected cells. By using a chromogenic β-galactosidase substrate, HSV-infected cells stain intensely blue whereas uninfected cells show no staining. Both HSV-1 and HSV-2 are detectable with this cell line, but after infection with HCMV, varicella zoster virus (VZV), adenovirus, and Sindbis virus, no blue cells are seen.

In the first clinical study of this technology, BHKICP6LacZ cells and the histochemical staining method compared well with the CPE assay in detecting HSV in clinical specimens submitted to a diagnostic virology laboratory.[50] In this study, a technologist with no prior experience in reading histochemical staining determined positive and negative staining at 16 to 24 hr after inoculation. Detection of CPE requires careful microscopic observation by a trained eye, often over a period of several days to a week before specimens are identified as positive or are confidently reported as negative. In addition, after CPE appears, most laboratories confirm the result with fluorescent antibody staining of the cells. The ability to report positives three or more days early using the histochemical method saves many hours of microscopic observation of tube cultures. Furthermore, being able to confidently report negative results within a day or two saves even more hours because the cultures would not need to be observed for seven days. This study, therefore, showed that the BHKICP6LacZ cell line is useful as a rapid, sensitive, and specific assay for detecting HSV in clinical specimens and has several advantages over traditional culture methods.

This basic methodology was recently adapted into a commercially available FDA approved kit (ELVIS™-HSV kit, Diagnostic Hybrids, Inc. (DHI), Athens, OH). The ELVIS™-HSV assay detects HSV-1 and HSV-2, and no antibodies are required unless typing is necessary. Initial clinical trials revealed that ELVIS™-HSV performed well compared to a CPE assay or a shell-vial assay for detecting HSV.[51,52] Recently, ELVIS™-HSV was compared with an immunofluorescence/shell vial assay for HSV.[53] Both tests have equivalent sensitivity and specificity, but ELVIS™-HSV detected all of the positives in one day whereas less than 40% of the positives were detected in one day by shell vial. These authors conclude that the ELVIS™-HSV assay is simple to perform, rapid, sensitive, and specific.

Automated methods have enabled clinical diagnostic laboratories to process large numbers of specimens rapidly and have had a dramatic impact on how clinical samples are processed. Therefore an obvious goal of this area of research is to develop an automated assay for infectious HSV with sufficient sensitivity to detect a single infectious particle. The luciferase assay is reported to be many times more sensitive than any of the β-galactosidase assays.[42] Recently a cell line (BHKICP6LucA6) was isolated in which the firefly luciferase gene was tested as a reporter gene for this purpose.[44] Infected BHKICP6LucA6 cells exhibited a signal-to-noise ratio that was of magnitude sufficient that measuring the luciferase activity of an infected-cell lysate detected a single infected-cell when a practical number of cells (e.g., 10^4 to 10^5) were used in the assay. These results suggest that an automated virus detection assay could be developed with sufficient sensitivity for use by clinical laboratories.

3.2.2b. Other Herpesviruses. It would be desirable to adapt this technology to other human herpesviruses, such as HCMV, VZV, EBV, and human herpesviruses type 6, 7, and 8 (HHV6, HHV7, and HHV8). To the extent that they have been studied, all herpesviruses have a lytic replicative cycle that is similar to HSV. Therefore, the same basic approach outlined previously for HSV may be directly adaptable to these viruses. However, unlike HSV which replicates rapidly to high levels in a wide variety of cells, HCMV, VZV, EBV, HHV6, HHV8, and HHV8 are highly fastidious, have a narrow host-cell range, and replicate relatively slowly. HCMV, for example replicates well only in primary cultures of diploid human cells, such as primary human embryonic lung cells (MRC5 cells, etc.). These cell lines are very difficult to use as parental cells for generating stably transformed cell lines, significantly hampering efforts to isolate a cell line for HCMV detection. HCMV is a significant human pathogen primarily for the immunocompromised patient, and its isolation and identification is taking up an increasing fraction of the effort of the clinical virology laboratory at tertiary care centers. More rapid and simpler methods to detect HCMV would, therefore, be valuable. The increasing problem of drug-resistant VZV in AIDS patients may increase the need to culture and perform drug susceptibility testing on VZV-detection cell lines. There is presently no clinically useful cell culture propagation system for EBV because it undergoes a predominantly latent infection of B lymphocytes *in vitro*. Therefore, sophisticated genetic modification is required to generate a permissive cell line for propagating and detecting EBV. Recent encouraging reports, mentioned before, describe cell lines that have been rendered at least partially permissive for EBV by engineering to stable express the EBV receptor.[34] Increasing knowledge of the molecular genetics of HHV6 and HHV7 should facilitate development of detection cell lines for these viruses and for HHV8 when systems to culture it become available.

3.2.3. *Positive-Strand RNA Viruses*

3.2.3a. A Prototypical Cell Line for Alphaviruses. Recently, a sensitive assay was reported for detecting alphaviruses and alphavirus replicons based on a cell line (BHKSINLuc2) that contains a cDNA of a defective Sindbis viral genome.[43] Sindbis virus is an enveloped, positive-strand RNA virus and is the prototypical member of the alphavirus genus of the *Togaviridae* family.[54] It is a minor human pathogen, but its life cycle is similar to other alphaviruses such as eastern and western equine encephalitis viruses, and rubella virus, a more distantly related togavirus. Significant advances in understanding the molecular events involved in replicating togaviruses have been cleverly applied to the development of viral expression vectors.[55] Cell lines that facilitate detecting low levels of these viruses should also provide useful research tools and in the future may prove valuable in diagnostic laboratories.

Sindbis virus contains a single strand of positive-sense RNA that functions as mRNA.[54,56] The 5′ two-thirds of the genome codes for the viral nonstructural proteins required for replication and transcription of the viral RNA. In infected cells, this segment of the genome is translated from a single ORF, and co- and posttranslational cleavage of the encoded polyprotein produces four polypeptides. The 3′ one-third of the genome codes for the viral structural proteins. These proteins are not translated from the viral genomic RNA but from a subgenomic RNA that is transcribed form the complimentary negative strand of the genomic RNA.[57] RNA transcription from this subgenomic promoter is very robust and this forms the basis for the Sindbis viral expression vectors.

BHKSINLuc2 cells were generated by using principles similar to those used to complement and package Sindbis replicons,[58] but the defective RNA used contains a reporter gene downstream of the subgenomic RNA promoter in place of the structural genes. The reporter gene, which codes for the enzyme luciferase, is expressed only when the cells are infected with Sindbis virus and synthesize the nonstructural proteins. BHKSINLuc2 cells exhibit high levels of luciferase activity after infection but essentially no activity beforehand above the baseline. Luciferase activity is detected as early as 4 h after infection, and luciferase activity at 6 to 8 hr post infection is proportional to the number of plaque-forming units (PFU) added, in the range of 10^4 to 10^5 PFU. Thus BHKSINLuc2 cells are permissive for Sindbis virus and provide a sensitive assay for detecting this virus. The cis-acting regulatory sequences, including the promoter for the subgenomic RNA, are highly conserved among alphaviruses[56,59] and, as expected, another alphavirus, Semliki Forest virus, induces luciferase activity in BHKSINLuc2 cells. Many other unrelated viruses do not trigger luciferase expression, but surprisingly, luciferase activity is induced by HSV, although the levels are several orders of magnitude lower than for Sindbis virus. Although this type of specificity problem must be solved, it is clear that BHKSINLus2 cells have set the stage for developing more clinically relevant cell lines based on this approach.

3.2.3b. Other Positive-Strand RNA Viruses. The strategy presented previously for Sindbis virus may be directly adaptable to viruses with similar replicative pathways, such as rubella virus.[60] Many positive-strand RNA viruses (e.g., enteroviruses and hepatitis A virus), however, do not produce a subgenomic RNA molecule. Their RNA genomes often contain a single ORF for translation, and viral proteins are produced by co- and posttranslational cleavage of a single polyprotein. It may be possible to alter the viral RNAs so that they contain a reporter gene in addition to the cis-acting sequences required for replication. A cell that expresses a sufficiently low amount of such a viral RNA might display a level of constitutive reporter gene expression below the level of detection. When cells are infected by the homologous virus, the defective RNA would be amplified resulting in a much increased level of translation of the reporter gene. An alternative approach would be to embed the reporter gene in a viral genome that is translated into a polyprotein. Cleavage of the polyprotein to release a functionally active reporter protein would require the action of a trans-acting protease provided by the homologous infectious virus. Increasing knowledge of such virally specified activities will make it feasible to test these strategies.

3.2.4. Negative-Strand RNA Viruses

Many of the viruses detected by the clinical virology laboratory are negative-strand RNA viruses (e.g., influenza viruses, parainfluenza viruses, RSV). Although negative-strand RNA viruses exhibit considerable diversity in genomic structure and biological properties, their RNA replicative and transcriptive strategies share common features. As with all RNA viruses, the replicative strategy of negative-strand RNA viruses is based on the fact that cells do not have the enzymes necessary to carry out RNA replication.

All RNA viruses encode an RNA-dependent RNA polymerase and other RNA replicase and transcriptase factors. Whereas positive-strand RNA viruses bring the mRNA for the RNA replicase into the cell and it is immediately translated after infection, negative-strand viruses have genomes that are opposite in polarity to their mRNA. Therefore, negative-strand viruses have a fundamentally different strategy to transcribe their mRNA and replicate their genome: they bring the RNA polymerase itself into the cell as part of the virion. Once the virion enters the cell and begins to uncoat, the virion RNA polymerase uses the genomic RNA as a template to transcribe mRNAs. The polymerase also replicates full-length, positive-strand, replicative-intermediate RNA which is used as template for making many copies of genomic RNA that are used then as templates for secondary transcription of additional viral mRNAs. The RNA polymerase is tightly associated with the virion ribonucleoprotein, and for this and other technical reasons, until very recently, the study of negative-strand viruses has not been amenable to direct genetic manipulation by recombinant DNA technology.

Recent methodological advances in the molecular manipulation of a number of negative-strand and double-strand RNA viruses have significantly improved virologist's ability to identify the specific mechanisms involved in replicating these viruses.[61] Using an approach first applied to influenza A virus, workers in this field have employed RNA derived from viral cDNA *in vitro* and a ribonucleoprotein transfection system, or they have generated viral RNA from cDNA in the cytoplasm using a T7 RNA polymerase system. These methods have been combined with reporter genes and/or defective interfering RNAs to assess RNA replication and transcription in virally-infected cells.[61–68]

These recent advances in the molecular manipulation of negative-strand RNA viruses, in addition to facilitating studies on the molecular mechanisms of viral replication and the development of expression vectors,[69] should facilitate the development of cell lines for their detection. Now it is conceivable that a cell line could be generated that constitutively expresses an RNA molecule containing a reporter gene ORF and the cis-acting sequences necessary for carrying out replication and transcription by the replicase and transcriptase of the same negative-strand RNA virus. As with the Sindbis virus case described before, this RNA would be constructed so that the reporter gene ORF is not translated in uninfected cells. Infection of this hypothetical cell line with the negative-strand RNA virus would bring the replicase/transcriptase complex into the cell. Then this complex would replicate and transcribe the resident ribonucleoprotein which encodes the reporter gene thus generating large amounts of translatable reporter gene mRNAs. As with the positive-strand RNA viruses, the specificity of the system would be conferred by the specific interaction of the viral RNA polymerase and its cofactors with ribonucleotide sequences on the chimeric RNA molecule.

4. FUTURE PROSPECTS

4.1. Engineering Cells Resistant to Viruses

There are certain circumstances in which one might want to eliminate the susceptibility of a cell to a particular virus. Certain contaminating viruses, such as the retrovirus monkey foamy virus, plague cell culture providers and users. Genetically modifying cells by deleting genes that encode factors essential for entry or replication of a contaminating virus, or by engineering cells to express an inhibitory factor, might provide a solution to this type of problem. In addition, modifying cells to restrict their susceptibility to certain viruses could help improve the specificity of particular detection cell lines.

4.2. Primary Cell Lines Derived from Transgenic Animals

Recent advances in the technology for generating transgenic animals have provided a powerful tool for studies of viral pathogenesis and for discerning the functional importance of certain genes and their products during viral replication in animals.[67–73] Theoretically, this same technology could be used to the generate primary cell lines for viral propagation and detection. Primary monkey, rabbit, and human cell lines continue to be an important part of the cell-line armamentarium of all clinical virology laboratories. Their importance could be reduced or eliminated in the future by genetically modifying continuous cell lines to behave like primary cell lines, at least from a virologist's perspective. Another possible approach, however, is to combine the technology described in the previous pages with transgenic animal technology. Then primary cell lines could be harvested from transgenic animals. These cells would have all of the virological benefits of primary cell lines and also the benefits of genetically modified cell lines for facilitating detection of viruses.[70] Cell lines could also be generated from animals transgenetically made resistant to a virus.[74,75]

5. CONCLUSION

Genetic engineering has held the promise of an exciting array of new products in areas as diverse as botany and medicine. Novel vaccines, therapeutic agents, and diagnostic assays based on the methods of genetic engineering will soon be commonplace. Transgenic cell lines and animals have already become standard tools of biomedical research and are now a reality in the clinical virology laboratory. It is possible that they will change the way diagnostic virology is performed in the not too distant future.

REFERENCES

1. Specter, S., and Lancz, G., 1992, *Clinical Virology Manual,* 2nd ed., Elsevier, Amsterdam.
2. Enders, J. F., Weller, T. H., and Robbins, F. C., 1949, Cultivation of the Lansing strain of poliomyelitis virus in cultures of various human embryonic tissues, *Science* **109:**85–87.
3. Dulbecco, R. and Vogy, M., 1953, Some problems of animal virology as studied by the plaque technique, *Cold Spring Harbor Symp Quant Biol.* **18:**273–279.
4. Hsiung, G. D., 1984, Diagnostic virology: From animals to automation, *Yale J. Biol. Med.* **57:**727–733.
5. Hsiung, G. D., 1989, The impact of cell culture sensitivity on rapid viral diagnosis: A historical perspective, *Yale J. Boil. Med.* **62:**79–88.
6. Landry, M. L., Mayo, D. R., and Hsiung, G. D., 1989, Rapid and accurate viral diagnosis, *Pharmacol. Ther.* **40:**287–328.
7. Shenk, T. E., 1993, Virus and cell: Determinants of tissue tropism, in: *Emerging Viruses* (S. S. Morse, eds.), Oxford University Press, New York, pp. 79–90.
8. Asanaka, M., and Lai, M. M., 1993, Cell fusion studies identified multiple cellular factors involved in mouse hepatitis virus entry, *Virology* **197:**732–741.
9. Bass, D. M., Baylor, M. R., Chen, C., Mackow, E. M., Bremont, M., and Greenberg, H. B., 1992, Liposome-mediated transfection of intact viral particles reveals that plasma membrane penetration determines permissivity of tissue culture cells to rotavirus, *J. Clin. Invest.* **90:**2313–2320.
10. Blair, W. S., Li, X., and Semler, B. L., 1993, A cellular cofactor facilitates efficient 3CD cleavage of the poliovirus P1 precursor, *J. Virol.* **67:**2336–2343.
11. Gotoh, B., Ogasawara, T., Toyoda, T., Inocencio, N. M., Hamaguchi, M., and Nagai, Y., 1990, An endoprotease homologous to the blood clotting factor X as a determinant of viral tropism in chick embryo, *EMBO J.* **9:**4189–4195.
12. Nagai, Y., Inocencio, N. M., and Gotoh, B., 1991, Paramyxovirus tropism dependent on host proteases activating the viral fusion glycoprotein [Review], *Behring Inst. Mitt.* **89:**35–45.
13. Eki, T., Enomoto, T., Masutani, C., Miyajima, A., Takada, R., Murakami, Y., Ohno, T., Hanaoka, F., and Ui, M., 1991, Mouse DNA primase plays the principal role in determination of permissiveness for polyomavirus DNA replication, *J. Virol.* **65:** 4874–4881.
14. Bruckner, A., Stadlbauer, F., Guarino, L. A., Brunahl, A., Schneider, C., Rehfuess, C. Previes, C., Fanning, E., and Nasheuer, H. P., 1995, The mouse DNA polymerase alpha-primase subunit p48 mediates species-specific replication of polyomavirus DNA in vitro, *Mol. Cell Biol.* **15:**1716–1724.
15. Yokomori, K., Asanaka, M., Stohlman, S. A., and Lai, M. M., 1993, A spike protein-dependent cellular factor other than the viral receptor is required for mouse hepatitis virus entry, *Virology* **196:**45–56.
16. De, B. P., Gupta, S. Gupta, S., and Banerjee, A. K., 1995, Cellular protein kinase C isoform ζ regulates human parainfluenza virus type 3 replication, *Proc. Natl. Acad. Sci.* USA **92:**5204–5208.
17. Norkin, L. C., 1995, Virus receptors: Implications for pathogenesis and the design of antiviral drugs, *Clin. Microbiol. Rev.* **8:**293–315.
18. Dragic, T., Charneau, P., Clavel, F., and Alizon, M., 1992, Complementation of murine cells for human immunodeficiency virus envelope/CD4-mediated fusion in human/murine heterokaryons, *J. Virol.* **66:**4794–4802.

19. Feigenbaum, L., Khalili, K., Major, E., and Khoury, G., 1987, Regulation of the host range of human papovirus JCV, *Proc. Natl. Acad. Sci.* USA **84:**3695–3698.
20. Martin, M. E., Piette, J., Yanvi, M., Tang, W. J., and Folk, W. R., 1988, Activation of the polymavirus enhancer by a murine activator protein 1 (AP1) homolog and two contiguous proteins, *Proc. Natl. Acad. Sci.* USA **85:**5839–5843.
21. Ori, A., and Shaul., 1995, Hepatitis B virus enhancer binds and is activated by the hepatocyte nuclear factor 3, *Virology* **207:**98–106.
22. Coen, D. M., 1991, Molecular genetics of animal viruses, in: *Fundamental Virology,* 3rd ed. (B. N. Fields and D. M. Knipe, eds.), Raven Press, New York, pp. 123–150.
23. Mendelsohn, C. L., Wimmer, E., and Racaniello, V. R., 1989, Cellular receptor for poliovirus: Molecular cloning, nucleotide sequence, and expression of a new member of the immunoglobulin superfamily, *Cell* **56:**855–865.
24. Pipkin, P. A., Wood, D. J., Racaniello, V. R., and Minor, P. D., 1993, Characterisation of L cells expressing the human poliovirus receptor for the specific detection of poliovirus in vitro, *J. Virol Methods* **41:**333–340.
25. Dalgleish, A. G., Beverely, P. C., Clapham, P. R., Crawford, D. H., Greaves, M. F., and Weiss, R. A., 1984, The CD4 (T4) antigen is an essential component of the receptor for the AIDS retrovirus, *Nature* **312:**763–767.
26. Klatzmann, D., Champagne, E., Charmaret, S., Gruest, J., Guetard, D., Hercend, T., Gluckman, J. C. and Montagnier, L., 1984, T-lymphocyte T4 molecule behaves as the receptor for human retrovirus LAV, *Nature* **312:**767–768.
27. Chesebro, B. and Wehrly, K., 1988, Development of a sensitive quantitative focal assay for human immunodeficiency virus infectivity, *J. Virol.* **62:**3779–3788.
28. Alkhatib, G., Combadiere, C., Broder, C. C., Feng, Y., Kennedy, P. E., Murphy, P. M., and Berger, E. A., 1996, CC CKR5: A Rantes, MIP-1α, MIP-1 β receptor as a fusion cofactor for macrophage-tropic HIV-1, *Science* **272:**1955–1958.
29. Choe, H., Farzan, M., Sun, Y., Sullivan, N., Rollins, B., Ponath, P. D., Wu, L., MacKay, C. R., LaRosa, G., Newman, W., Gerald, N., Gerald, C., and Sodroski, J., 1996, The β-chemokine receptors CCR3 and CCR5 facilitate infection by primary isolates of HIV-1, *Cell* **85:**1135–1148.
30. Deng, D., Liu, R., Ellmeier, W., Choe, S., Unutmaz, D., Burkhart, M., DiMarzio, P., Marmon, S., Sutton, R. E., Hill, C. M., Davis, B., Peiper, S. C., Schall, T. J., Littman, D. R., and Landau, N. R., 1996, Identification of a major co-receptor for primary isolates of HIV-1, *Nature* **381:**661–666.
31. Fingeroth, J. D., Weis, J. J., Tedder, T. F., Strominger, J. L., Biro, P. A., and Fearon, D. T., 1984, Epstein–Barr virus receptor of human B lymphocytes is the C3d receptor CR2, *Proc. Natl. Acad. Sci.* USA **81:**4510–4514.
32. Aherarn, J. M., Hayward, S. D., Hichey, J. C., and Fearon, D. T., 1988, Epstein–Barr virus (EBV) infection of murine L cells expressing recombinant human EBV/C3d receptor, *Proc. Natl. Acad. Sci.* USA **85:**9307–9311.
33. Crawford, D. H., 1994, Epstein–Barr Virus, in: *Principles and Practice of Clinical Virology,* 3rd ed. (A. J. Zuckerman, J. E. Banatvala, and J. R. Pattison, eds.), John Wiley & Sons, Chichester, pp. 109–143.
34. Li, Q. X., Young, L. S., Niedobitek, G., Dawson, C. W., Birkenbach, M., Wang, F., and Rickinson, A. B., 1992, Epstein–Barr virus infection and replication in a human epithelial cell system, *Nature* **356:**347–350.
35. Baltimore, D., 1971, Expression of animal virus genomes, *Bacteriol. Rev.* **35:**235–241.
36. Felber, B. K. and Pavlakis, G. N., 1988, A quantitative bioassay for HIV-1 based on trans-activation, *Science* **239:**184-187.
37. Akrigg, A., Wilkinson, G. W., Angliss, S., and Greenaway., P. J., 1991, HIV-1 indicator cell lines, *AIDS* **5:**153–158.

38. Kimpton, J. and Emerman, M., 1992, Detection of replication-competent and pseudotyped human immunodeficiency virus with a sensitive cell line on the basis of activation of an integrated β-galactosidase gene, *J. Virol.* **66:**2232–2239.
39. Rocancourt, D., Bonnerot, C., Jouin, H., Emerman, M., and Nicholas, J. F., 1990, Activation of a β-galactodisase recombinant provirus: Application of titration of human immunodeficiency virus (HIV) and HIV-infected cells, *J. Virol.* **64:**2660–2668.
40. Stabell, E. C. and Olivo, P. D., 1992, Isolation of a cell line for rapid and sensitive histochemical assay for the detection of herpes simplex virus, *J. Virolog. Methods* **38:**195–204.
41. Alam, J. and Cook, J. L., 1990, Reporter genes: Application to the study of mammalian gene transcription, *Anal. Biochem.* **188:**245–254.
42. de Wet, J. R., Wood, K. V., DeLuca, M., Helinski, D. R., and Subramani, S., 1987, Firefly luciferase gene: Structure and expression in mammalian cells, *Mol. Cell. Biol.* **7:**725–737.
43. Olivo, P. D., Frolov, I., and Schlesinger, S., 1994, A cell line that expresses a reporter gene in response to infection by Sindbis virus: A prototype for detection of positive strand RNA viruses, *Virology* **198:**381–384.
44. Olivo, P. D., 1994, Detection of herpes simplex virus by measurement of luciferase activity in an infected-cell lysate, *J. Virol. Methods* **47:**117–128.
45. Zolotukhin, S., Potter, M., Hauswirth, W. W., Guy, J., and Muzyczka, N., 1996, A "humanized" green fluorescent protein cDNA adapted for high-level expression in mammalian cells, *J. Virol.* **70:**4646–4654.
46. Mosca, J. D., Bednarik, D. P., Raj, N. B. K., Rosen, C. A., Sodroski, J. G., Haseltine, W. A., Haywood, G. S., and Pitha, P. M., 1987, Activation of human immunodeficiency virus by herpesvirus infection: Identification of a region within the long terminal repeat that responds to a transacting factor encoded by herpes simplex virus 1, *Proc. Natl. Acad. Sci.* USA **84:**7408–7412.
47. Mosca, J. D., Bednarik, D. P., Raj, N. B. K., Rosen, C. A., Sodroski, J. G., Haseltine, W. A., and Pitha, P. M., 1987, Herpes simplex virus type-1 can reactivate transcription of latent human immunodeficiency virus, *Nature* **325:**67–70.
48. Popik, W. and Pitha, P. M., 1991, Inhibition by interferon of herpes simplex virus type 1-activated transcription of *tat*-defective provirus, *Proc. Natl. Acad. Sci.* USA **88:**9573–9577.
49. Savatier, N., Rocancourt, D., Bonnerot, C., and Nicolas, J. F., 1989, A novel system for screening antiretroviral agents, *J. Virol. Methods* **26:**229–235.
50. Stabell, E. C., O'Rourke, S. R., Storch, G. A., and Olivo, P. D., 1993, Evaluation of a genetically engineered cell line and a histochemical β-galactosidase assay to detect herpes simplex virus in clinical specimens, *J. Clin. Microbiol.* **31:**2796–2798.
51. Hodinka, R. L., and Stetser, R. L., 1994, Evaluation of a commercial Enzyme-Linked Virus Inducible System (ELVIS) for the rapid detection of herpes simplex virus in clinical specimens, *10th Annual Clinical Virology Symposium*, Clearwater, FL., abstract No. T22, p.147.
52. Williams, B., 1994, Evaluation of the Enzyme-Linked Virus Inducible System (ELVIS) for the rapid detection of herpes simplex virus, *10th Annual Clinical Virology Symposium*, Clearwater, FL., abstract No. T21, p.146.
53. Proffitt, M. R. and Schindler, S. A., 1995, Rapid detection of HSV with an Enzyme-Linked Virus Inducible System (ELVIS (TM) employing a genetically modified cell line, *Clin. Diagn. Virol.* **4:**175–182.
54. Schlesinger, S. and Schlesinger, M., 1996, Togaviridae: The viruses and their replication, in: *Fundamental Virology*, 3rd ed. (B. N. Feilds, D. M. Knipe, and P. M. Howley, eds.), Lippincott-Raven, Philadelphia, pp. 523–539.
55. Bredenbeek, P. J. and Rice, C. M., 1992, Animal RNA virus expression systems, *Sem. Virol.* **3:**297–310.
56. Strauss, E. G. and Strauss, J. H., 1986, Structure and replication of the alphavirus genome, in:

The Togviridae and Flaviviridae, (S. Schlesinger and M. J. Schlesinger, eds.) Plenum Press, New York, pp. 35–82.
57. Levis, R., Schlesinger, S., and Huang, H. V., 1990, Promoter for Sindbis virus RNA-dependent subgenomic RNA transcription, *J. Virol.* **64:**1726–1733.
58. Bredenbeek, P. J., Frolov, I., Rice, C. M., and Schlesinger, S., 1993, Sindbis virus expression vectors: Packaging of RNA replicons by using defective helper RNAs, *J. Virol.* **67:**6439–6446.
59. Hertz, J. M., and Huang, H. V., 1992, Utilization of heterologous alphavirus junction sequences as promoters by Sindbis virus, *J. Virol.* **66:**857–864.
60. Wang, C. Y., Dominguez, G., and Frey, T. K., 1994, Construction of rubella virus genome-length cDNA clones and synthesis of infectious RNA transcripts, *J. Virol.* **68:**3550–3557.
61. Palease, P., 1995, Genetic engineering of infectious negative-strand RNA viruses, *Trends Microbiol.* **3:**123–125.
62. Enami, M., Luytejs, W., Krystal, M., and Palease, P., 1990, Introduction of site-specific mutations into the genome of influenza virus, *Proc. Natl. Acad. Sci.* USA **87:**3802–3805.
63. Park, K. H., Huang, T., Correia, F. F., and Krystal, M., 1991, Rescue of a foreign gene by Sendai virus, *Proc. Natl. Acad. Sci.* USA **88:**5537–5541.
64. Calain, P., Curran, J., Kolakofsky, D., and Roux, L., 1992, Molecular cloning of natural paramyxovirus copy-back defective interfering RNAs and their expression from DNA, *Virology* **191:**62–71.
65. Collins, P. L., Mink, M. A., and Stec, D. S., 1991, Rescue of synthetic analogs of respiratory syncytial virus genomic RNA and effect of truncations and mutations on the expression of a foreign reporter gene, *Proc. Natl. Acad. Sci.* USA **88:**9663–9667.
66. Gorziglia, M. I. and Collins, P. L., 1992, Intracellular amplification and expression of a synthetic analog of rotavirus genomic RNA bearing a foreign marker gene: Mapping cis-acting nucleotides in the 3'-noncoding region, *Proc. Natl. Acad. Sci.* USA **89:**5784–5788.
67. Yu, Q. Z., Hardy, R. W., and Wertz, G. W., 1995, Functional cDNA clones of the human respiratory syncytial (RS) virus N, P, and L proteins support replication of RS virus genomic RNA analogs and define minimal trans-acting requirements for RNA replication, *J. Virol.* **69:**2412–2419.
68. Pattnaik, A. K., Ball, L. A., Legrone, A., and Wertz, G. W., 1995, The termini of VSV DI particle RNAs are sufficient to signal RNA encapsidation, replication, and budding to generate infectious particles, *Virology* **206:**760–764.
69. Percy, N., Barclay, W. S., Garcia-Sastre, A., and Palease P., 1994, Expression of a foreign protein by influenza A virus, *J. Virol.* **68:**4486–4492.
70. Federspiel, M. J., Bates, P., Young, J. A., Varmus, H. E., and Hughes, S. H., 1994, A system for tissue-specific gene targeting: Transgenic mice susceptible to subgroup A avian leukosis virus-based retroviral vectors, *Proc. Natl. Acad. Sci.* USA **91:**11241–11245.
71. Racaniello, V. R., Ren, R., and Bouchard, M., 1993, Poliovirus attenuation and pathogenesis in a transgenic mouse model for poliomyelitis, *Dev. Biol. Stand.* **78:**109–116.
72. Ren, R. B., Costantini, F., Gorgacz, E. J., Lee, J. J., and Racaniello, V. R., 1990, Transgenic mice expressing a human poliovirus receptor: A new model for poliomyelitis, *Cell.* **63:**353–362.
73. Ren, R., and Racaniello, V. R., 1992, Human poliovirus receptor gene expression and poliovirus tissue tropism in transgenic mice, *J. Virol.* **66:**296–304.
74. Han, L., Yun, J. S., and Wagner, T. E., 1991, Inhibition of Moloney murine leukemia virus-induced leukemia in transgenic mice expressing antisense RNA complementary to the retroviral packaging sequences. *Proc. Natl. Acad. Sci.* USA **88:**4313–4317.
75. Pavlivic, J., Arzet, H. A., Hefti, H. P., Frese, M., Rost, D., Ernst, B., Kolb, E., Staeheli, P., and Haller, O., 1995, Enhanced virus resistance of transgenic mice expressing human MxA protein, *J. Virol.* **69:**4506–4510.

2

Use of Synthetic Peptides in Microbial Diagnostics

SUSANNE MODROW and HANS WOLF

1. INTRODUCTION

Since the introduction of solid-phase peptide synthesis by Bruce Merrifield[1] in 1963, chemical methods for synthesis have been considerably improved by using milder conditions.[2] The introduction of automated systems together with further methods developed during the past few years simplified the synthesis.[3–6] Therefore today it is possible for peptide chemists and also for scientists with their primary experience in molecular biology, genetics, infectious diseases, or immunology to employ a variety of synthetic techniques and to synthesize up to thousands of various peptides in appropriate quantities and purities at relatively reasonable costs compared with previous methods. As peptides have become readily available, they have found widespread use as reagents in natural sciences and medicine.

Synthetic peptides are employed as B cell epitopes, which are specifically recognized via the variable domain of immunoglobulins. Alternatively oligopeptides stimulate the cellular immune system because they are incorporated into the antigen-binding groove of MHC-class I or MHC-class II proteins present at the surfaces of antigen-presenting cells and interact with T cell receptor complexes of cytotoxic T lymphocytes or T helper cells, respectively. In the first case these

SUSANNE MODROW AND HANS WOLF • Institute for Medical Microbiology and Hygiene, University of Regensburg, D-93053 Regensburg, Germany.

Rapid Detection of Infectious Agents, edited by Specter *et al.* Plenum Press, New York, 1998.

processes induce the elimination of the antigen-presenting cells by lysis, with respect to the second example, the secretion of cytokines is induced and is followed by T cell proliferation. Synthetic peptides for diagnostic test systems have been introduced, in particular, as specific antigens for detecting antibodies in patient's pathological sera that are directed against protein components of infectious agents, be they viral,[7,8] bacterial,[9] or parasitic.[10] In combination with the highly developed automated systems for nucleic acid amplification and DNA sequencing, synthetic peptides provide easily and rapidly available antigens where the genetic information of a recently isolated new infectious agent has been characterized. The easy and fast distinction between infections caused by different viral subtypes, for example, human immunodeficiency virus (HIV),[11] human T cell lymphotropic virus (HTLV),[12] and respiratory syncytial viruses (RSV)[13] was shown as another major advantage of diagnostic tests based on synthetic peptides.

2. METHODOLOGY

2.1. Peptide Synthesis

Oligopeptides are generally synthesized by coupling with the α-carboxylic acid group of the peptide's carboxyterminal residue to the solid phase, which may be represented by a matrix consisting of polystyrene,[1] polystyrene-polyoxyethylene graft copolymers,[14] cellulose,[15] derivatized silica substrates,[16] or other materials. Depending on the applied methodology, the matrices have different forms, such as small beads (resins), filter discs, pins, or sticks. The peptide chains are built by consecutively adding the respective amino acids protected at the amino group either by *t*-Boc (*t*-butyloxycarbonyl) or Fmoc (flourenylmethyloxycarbonyl) groups, which are removed by applying TFA or piperidine, respectively, after the successfully completed coupling reaction. These processes of coupling and deprotection are repeated for every residue to be added to the growing peptide chain. In a final step the peptide has to be cleaved from the solid phase matrices by either HF or trifluormethane sulfoic acid (TMSFA) treatment in cases when *t*-Boc-protected amino acids have been used during synthesis or by incubation with TFA when Fmoc-protected residues have been applied. Simultaneously the chemical groups used for protecting the amino acid's functional groups at the β-side chains during synthesis are generally removed. The deprotected peptide is precipitated from solution and purified by reversed phase HPLC. The correct sequence is checked by amino acid sequencing, mass spectroscopy, or other methods.

The description given for the process of solid-phase peptide synthesis is a general outline. It has to be adjusted with respect to the amino acid composition and the sequence of the individual peptide. The methods and the materials which have to be applied in special cases have been thoroughly reviewed, and if readers want to use solid-phase peptide synthesis, they are advised to consult the respective books and overviews.

In addition to the method of synthesizing and characterizing one individual peptide, various methods for multiple peptide synthesis have been developed. These approaches are especially useful for laboratories whose research requires large numbers of synthetic peptides. The PEPSCAN method had been developed for screening a protein sequence for antigenic B and T cell epitopes.[3,4,17,18] Oligopeptides about 9 to 12 residues long are synthesized by using polystyrene pins as solid supports. The pins are arranged so that they fit into a 96-well ELISA plate. Usually a complete set of peptides overlapping in one residue is prepared to cover the whole sequence of the protein of interest. After cleavage of side-chain protection groups, the peptides remain covalently bound to the pins and are directly incubated with patients' sera to identify the antigenic domains. The peptide-bound pins are cleaned to remove antibodies by SDS/mercaptoethanol treatment and are reused several times for testing additional sera.

More than one hundred different peptides can be synthesized simultaneously using the "tea bag" method.[5] In this case solid supports in the form of resins or filter discs[16] with the growing amino acid chains are used and sealed in polystyrene bags into which solvents and chemicals can diffuse. Similarly as with the pin technology, this method has the advantage that an entire set of different peptides may be synthesized using common washing and deprotection steps. Only the addition of the specific amino acid requires an individual treatment. Using the tea bag technology, it is possible to synthesize milligram quantities (instead of microgram quantities with the pin method) of the individual peptides which can be cleaved from the matrix, purified, further characterized, and applied as free peptides.

An extension to these methods for multiple peptide synthesis is offered by the recently introduced peptide libraries, which are composed of tens to hundreds of millions of different sequences.[20] They provide a general and broad approach for identifying novel peptides when only limited knowledge is available concerning the amino acid sequence parameters necessary for using a peptide or protein domain in diagnostic or pharmaceutical assays. Peptide libraries are generated by a "physical" approach based on resin mixing, that means dividing the resin batch, coupling the amino acids, and recombing the resins with the newly added residues.[21,22] This method produces almost equimolar concentrations of each potential peptide. It is limited, however, to peptides about five residues long.[23] The second approach for generating a peptide library is termed the "chemical" synthesis method. In this case the peptide mixture resins representing the library are prepared from a specific ratio of amino acids, empirically defined to give an equimolar incorporation of each residue at each coupling step.[24–26] The format or configuration of a peptide library with defined amino acids at two or more positions is one of the most important aspects to be considered. Another important precondition for using peptide libraries is the availability of an effective screening protocol for identifying reactive peptides, e.g., those representing an antigenic determinant in ELISA assays. Only where those test systems exist is the reactive "leading sequence" identified from a library's peptide mixture.

As an alternative to those methods for multiple peptide synthesis, libraries of peptides displayed by filamentous bacteriophages have been developed.[27] With this approach, also known as reverse genetics, random peptide sequences are produced as part of the phage-coat proteins and are displayed at the surface of the infectious particles. Similar to synthetic peptide libraries, random sequences displayed by bacteriophages represent a technology for identifying "leading sequences," e.g., for cross-reacting antigenic domains in cellular proteins when effective screening protocols are available.

2.2. Peptide Coupling

Generally, peptide antigens used in diagnostic assays for detecting specifically binding antibodies have to be attached to the solid phase of the perspective test system, in most cases to the wells of the ELISA-plates consisting of coated polystyrene. The easiest method is to directly absorb the purified peptides dissolved in 0.2 M carbonate buffer (pH 9.5) to the wells during an overnight incubation period. The majority of peptides, especially when residues with aromatic ring systems are part of the sequence, attach very efficiently under those conditions and amounts of 50 to 400ng per well are sufficient for a positive reaction. Problems arise especially when rather short peptides up to 10 residues long are used as antigens. Here covalent coupling to the ELISA plates, e.g., via gultardialdehyde, is required. Many companies offer plates precoated with glutardialdehyde and ready for use in their assortment. Another possibility is covalently coupling the peptides to carrier proteins which should not, of course, be immunologically recognized by the antibodies used in the test systems. Therefore, in assays used for detecting immunogloblins in human sera HSA (human serum albumin) may be employed. Different protocols using various substances for linking the peptide to the carrier protein are available and depend of the presence of reactive amino acids in the peptide sequence. Frequently used are glutardialdehyde,[29] carbodiimide,[30] and *N*-maleimido-succinimidesters.[31] In every case the coupling has a major influence on the presentation of the peptide antigen and therefore on the binding and the reactivity of the antibody. It is generally to that of the native protein should be selected. Covalent linkage to the carrier protein or to the wells of the ELISA plates should be done preferable via the carboxy- or amino-terminal residues. One possibility is coupling via the SH-group of an additionally included cysteine, which removes the need to use the carboxy or amino funcıions of internal amino acids (Glu, Asp, Lys). These residues are often important for antigenicity, and the corresponding antigenic sites may be destroyed when coupling agents, such as carbodiimides or glutardialdyde, are used.

An alternative to coupling the purified peptides to the solid support of the test system is the use of the same polystyrene carrier used during peptide synthesis also as support for the antibody reaction in ELISA assays. The PEPSCAN technology described previously as a method for multiple peptide synthesis uses polystyrene pins as a solid phase for peptide synthesis and also for peptide presentation to antibodies.[3,4,17] Similar methods have been developed for synthesizing peptides on polystyrene resins or beads.[22,32,33] The peptides remain covalently bound to the matrices via their carboxy-terminal ends. Because purification is not possible, every amino acid coupling step has to be optimized to avoid the presence of truncated peptides or variants with individual residues lacking in the sequence.

2.3. Identification of Antigenic Epitopes

Before using synthetic peptides as antigens in diagnostic assays, it is necessary to identify and characterize the major antigenic epitopes in a protein's sequence which are regularly to be recognized by antibodies in patients' sera. The smallest B cell epitopes have been described as four to six amino acids.[4,34] When applied in ELISA test systems for detecting specific immunoglobulins generally peptides from 12 to 35 amino acids long are used. The shorter sequences, in particular, represent preferentially sequential epitopes because structural epitopes formed by folding the amino acid chain into a secondary or tertiary structure cannot be simulated by short peptides in the same way as they are formed in the native protein. During the last few years, two main procedures frequently used for identifying B-cell epitopes have been developed.

The first procedure is based on empirical methods as the synthesis of overlapping peptides spanning the entire protein sequence,[3,4,17,35–38] the selection of the preferentially recognized epitope as part of a synthetic peptide library by immunological screening[21,39], or the use of peptide fragments by proteolytic cleavage of the purified polypeptide, and testing these fragments for reactivity with sera from infected individuals.[40,41] In contrast to epitopes recognized by the humoral immune system, T cell epitopes fitting into the antigen-binding groove of MHC-class-I antigens have a defined length of about 8 to 10 amino acids,[42] determinants binding to MHC-class-II antigens comprise up to 20 residues.[43] The empirical approach using overlapping synthetic peptides that span the total sequence of the protein in question are also applied to characterize T cell epitopes.[44–46]

The second way of identifying antigenic epitopes is based on a predictive approach. The optimal way to predict B cell epitopes is based on the knowledge of the amino acid sequence of a protein and also on structural parameters. When the protein structure is known by X-ray crystallography or NMR analysis, the areas of amino acids on the surface of the individual protein molecule are determined. Once the three-dimensional structure is known, it is possible to predict sequential and also discontinuous, structurally defined antigenic sites. Those ideal conditions are very rare. In most cases only the genetic information about an infectious agent and the proteins encoded is available with limited knowledge on protein function or structural parameters. Therefore, in most cases it becomes necessary to predict anitgenic epitopes from the primary amino acid sequence of a protein.

Because antibodies produced in the course of an infection interact with structurally and functionally active proteins, they preferentially interact with regions on the surface of the infectious agents or the proteins, respectively. Amino acids in the interior part of a protein or of the protein core are accessible only to antibody binding when structure is disrupted, e.g., by treatment with detergents or chaotropic agents. In these cases even protein areas which are not regularly exposed become accessible for antibody binding. This is, however, not possible during normal epitope recognition by antibodies, whose production is induced during the course of an infection and which are of diagnostic relevance. Here, immunoglobulins preferentially interact with surface-exposed protein domains. Those regions are in the majority determined by a special subset of amino acids which renders them predictable as potentially antigenic sites. Beside other values the local hydrophilicity,[47,48] atomic mobility,[49,50] and flexibility[51] contribute to the surface accessibility of a protein domain.[52,53] An additional factor is the amino acid variability, when sequences of several functionally identical proteins, e.g., in related species, are known.[54,55] These parameters do not represent independent variables but are generally related to each other. Protein regions combining several or all of those parameters are present in surface projections and form exposed loops. In addition to these factors which depend mainly on the chemical properties of the individual amino acids, features of the structural propensities contribute to the antigencity of a protein region. Based on the known three-dimensional structures of a small number of polypeptides, the probability of secondary structures is calculated from the amino acid sequence. Certain motives, such as β-turns are common for antigenic sites.[56,57] Therefore, it is possible to predict continuous B cell epitopes by analyzing the amino acid sequence of a protein by using computer programs that link theoretical and empirical data of the chemical properties of the individual residues with structural conformational parameters.[58] This method of predicting epitopes has been and is successfully applied to a large number of proteins encoded by various organisms.[59]

In contrast to B cell epitopes, which are recognized as defined parts or structures of native proteins, T cell epitopes are represented by short peptides derived from the individual polypeptides by proteolytic cleavage. A very important contribution for rationally predicting T cell determinants came from the publication of the crystallographic structures of MHC-class-I and -class-II antigens.[43,60,61] It became obvious that MHC molecules bind peptides derived from cellular or foreign (viral, bacterial, parasital) proteins in a groove formed by two α-helices topped onto a β-pleated-sheet plateau. Pockets were identified at the bottom of the groove into which mainly hydrophobic residues at defined positions of the peptide can be fitted.[62] By analyzing different MHC subtypes and by eluting peptides complexed with various MHC subtypes it could be shown that the location of the pockets and the distance separating them are slightly different. Therefore the hydrophobic residues representing the anchors have to be present at defined positions to enable the respective peptides to interact with a specific MHC subtype. Commonly used algorithms for sequence homology or pattern recognition are used to find the respective motifs by analyzing the primary amino acid sequence of a protein and to predict peptides that specifically form complexes with a defined MHC-subtype molecule.[63,64]

3. APPLICATIONS

3.1. Research

Synthetic peptides have been used in nearly all fields of natural sciences and medicine, and it seems impossible to summarize all of them. Oligopeptides, in particular, have proven to be useful tools in studying the functional activities of proteins in immunology and in the development of pharmacologically relevant compounds. In immunology, synthetic peptides were applied to identifying antigenic domains in viral, bacterial, and protozoal proteins. Such peptides were helpful in characterizing the immune response developed against foreign agents and antigens and the immunological reactions directed to self-anitgens in autoimmune diseases. Primarily, when new infectious agents have been isolated and the nucleic acid sequence has been determined by sequence analysis, synthetic peptides allow the rapidly characterizing proteins and epitopes which are diagnostically important for detecting specific immunglobulins in infected individuals. It has to be considered, however, that eptiopes mapped with PEPSCAN technology or by synthetic peptide libraries result in the identifying relatively short amino acid sequences which may be used as "leading sequences" to develop an optimal peptide antigen for diagnostic teat systems. It has been shown that short peptides up to 12 residues long are highly flexible and interact nonspecifically with the variable domains of immuoglobulin molecules. To avoid these reactions that lead to false positive test results, additional amino acid sequences flanking the epitope have to be included in the antigen's sequence and result in developing a synthetic peptide that binds antibodies with high affinity and specificity.

Synthetic peptides derived from the amino acids sequence of many proteins were and are used to induce poly- or monoclonal antibodies which may be used generally to characterize the respective proteins and to analyze the function and the antigenicity of individual protein domains and the capacity of the peptide-specific antibodies to neutralize the respective infectious agents. Based on these studies, synthetic peptides may be used as synthetic vaccines to induce immune responses which protect against the infections and the diseases caused by viruses, bacteria, bacterial toxins or parasites. Even if the application of synthetic peptides as vaccines in humans is still hampered by the lack of appropriate immunological adjuvants, peptide compounds are very important research reagents in vaccine development to identify and to characterize relevant B and T cell epitopes.

Even if the direct application of unmodified synthetic peptides for treating diseases is unrealistic because the short half-life of the compounds in natural systems, peptides also represent important "leading structures" for developing immune modulators,[65] antivirals,[66] antibiotics,[67] and enzyme inhibitors[68]—to mention only some of the potential applications. Here, peptides are the molecular basis for designing new compounds that combine optimal functional activities with low degrees of unwanted side reactions. Even if the initially identified peptide shows rather poor biological activity, it may be manipulated by chemical modifications and further refinement studies to finally give the desired property profile.

3.2. Routine Diagnostics

Serodiagnosis of most infections is done routinely in ELISA test systems using preparations of the infectious agents isolated from biological material or culture systems or purified preparations of individual proteins specific for the respective virus, bacterium, or parasite may be used. The necessary polypeptides are isolated from culture systems used to propagate the agent or they are produced by recombinant gene technology, e.g., in *E. coli, Saccharomyces cerevisiae,* baculoviruses, and are purified. Alternatively, peptides derived from the amino acid sequence of a diagnostically relevant protein is chemically synthesized and used as antigen in ELISA assays. All of those methods have their specific advantages and disadvantages: the techniques associated with the propagating and manipulating high concentrations of highly infectious and sometimes dangerous agents are very expensive and time-consuming. Often appropriate cell culture systems are not available and the respective agents may be isolated only from the blood or stool of the infected patients or from infected lab animals. When proteins are produced by recombinant DNA technology, the antigen preparations have to be highly purified before application in ELISA to avoid false positive results caused by contamination with cellular or bacterial cross-reacting proteins. This high level of purity is often difficult to achieve and requires extensive research. The antigen, however, may be inexpensively produced in almost unlimited quantities, and the expressed regions may span the whole sequence of the respective protein including all of the desired structural and sequential epitopes.

Synthetic peptides represent a third possibility for producing diagnostically relevant antigens. As preconditions, the amino acid sequence of the respective protein and some knowledge of the location of the relevant epitopes must be available. As described previously, automated systems and improved chemical methods allow the synthesizing peptides in a releatevely short time and at low cost in almost any laboratory. The products are free of contaminating proteins and can be used in specific ELISA tests after a one-step column purification. They represent small protein domains that span only a limited number of epitopes. In most cases one peptide contains only one single epitope. Despite progress in chemical methods, the synthesizing longer peptides is still rather difficult and expensive.

As a consequence, synthetic peptides used in routine diagnosis for detecting immunoglobulins in patients' sera or mucosal secretions have to be applied as sets to avoid false negative results, as is done in a synthetic peptide multiple antigen test used to detect antibodies directed against proteins of HIV, Epstein–Barr virus, hepatitis C virus or *Leishmania donovani*.[10] Table I gives an overview of the commercially available test systems that use synthetic peptides as antigens. In some applications synthetic peptides are used in mixtures together with protein antigens produced by recombinant gene technology. Alternatively several epitopes have to be connected by amino acid spacers.[69] Especially when diseases caused by infectious agents with a high tendency to genetic variation are serologically diagnosed, ELISA tests based on synthetic peptides have been described occasionally as inefficient to detect antibodies directed against special subtypes.[70,71] Today ELISA assays based on synthetic peptides are used mainly for the initial and rapid identification of diagnostically relevant epitopes for special diagnostic problems. For instance, a synthetic peptide derived form a protein encoded by *Echinococcus granulosus*, a helminth which cannot be grown or amplified by available methods, proved an important antigen for detecting specific antibodies in infected individuals.[72] Generally, however, the serological differentiation between similar subtypes of the same infectious agents, particularly of virus subtypes, is one of the main fields where synthetic peptides represent optimal antigens. In these cases synthetic peptides have been applied in epidemiological studies to differentiate between the subtypes of HIV[11,33], or HTLV.[12,73] These peptide-based differentiation ELISAs are applied mostly to sera that have been confirmed positive by other test systems for antibodies against the respective viruses. However, synthetic peptides offer an easy and rapid possibility and are used in routine diagnostic assays for supplemental testing and for differentiation,

TABLE I
Licensed Diagnostic Test Systems Containing Synthetic Peptide Antigens Used for Antibody Detection in Human Sera

Name of test system[a]	Virus/ system	Synthetic peptides exclusively	Synthetic peptides combined with recombinant proteins	Company
Abbott Plus EIA 3rd generation	HIV screening		+	Abbott Diagnostika
Abbott Prism	HIV screening		+	Abbott Diagnostika
AxSym reagent pack	HIV screening		+	Abbott Diagnostika
HIV AB HIV-1/HIV-1 (rDNA) EIA	HIV screening		+	Abbott Diagnostika
IMx HIV-1/HIV-2 III plus reagent pack	HIV screening		+	Abbott Diagnostika
Enzygnost Anti HIV 1/2 plus	HIV screening		+	Behring
Enzygnost Anti HIV-2	HIV-1/HIV-2 screening	+		Behring
Enzymun-Test Anti-HIV 1+2+subtype 0	HIV screening		+	Boehringer Mannheim
Cobas Core Anti-HIV-1/2 EIA Dags	HIV screening		+	Hoffmann-LaRoche
ICE* HIV-1.0.2	HIV screening		+	Murex Diagnostika
Murex anti-HIV-1	HIV screening		+	Murex Diagnostika

Murex anti-HIV-1 and -2	HIV-1/HIV-2 screening		+	Murex Diagnostica
Vironostika HIV Uni-From II	HIV screening		+	Organon Teknika
LiaTek HIV-1+2	HIV-1/HIV-2 differentiation		+	Organon Teknika
Genelavia Mixt	HIV screening		+	Sanofi Diagnostics Pasteur
Genscreen HIV-1/2	HIV screening		+	Sanofi Diagnostics Pasteur
Pepti Lav 1-2	HIV-1/HIV-2 differentiation	+		Sanofi Diagnostics Pasteur
ETI-AB-HIV 1/2K	HIV screening		+	Sorin Biomedica
Inno Lia HIV antibody	HIV-1/HIV-2 differentiation		+	Sorin Biomedica
HIV Blot 2.2	HIV confirmation	+		Biochrom
ImmunoComb II HIV 1&2 BiSpot	HIV rapid testing	+		Hain Diagnostika
Enzymun-Test Anti-HCV	Hepatitis C virus screening		+	Boehringer Mannheim
Murex anti-HCV (version III)	Hepatitis C virus screening		+	Murex Diagnostika
ETI-AB-HCVK-3	Hepatitis C virus screening		+	Sorin Biomedica
Monolisa anti-HCV (new antigens)	Hepatitis C virus screening		+	Sanofi Diagnostics Pasteur
Inno-Lia HCV Antibody III	Hepatitis C virus screening		+	Innogenetics
Ortho anti-HCV (version III)	Hepatitis C virus screening		+	Ortho-Clinical Diagnostics
Chiron RIBA HCV 3.0 SIA	Hepatitis C virus confirmation		+	Ortho-Clinical Diagnostics

[a]Normally printed test systems licensed in Germany, bold printed test systems licensed in USA.

e.g., between HIV-1 and HIV-2[74] or HTLV-I and HTLV-II[70] and for subtyping different HIV-1 isolates.

3.3. Future Prospects

Today there are a variety of approaches for examining antibody detection tests in formats different from the routinely used ELISA test systems. The main goal is to develop an assay that is ready for use and reliable in both laboratory and non laboratory conditions. These methods include the use of peptides coupled to polystyrene particles, latex particles, or paramagnetic microparticles.[32,75] Such reagents can be used to develop very rapid and simple test systems. Another methodology based on synthetic peptides coupled via an Fab-fragment to the surface of red blood cells is highly specific for detecting HIV-specific antibodies in whole blood samples.[76] Another proposal is that various test systems based on biosensor or chip-array technology will offer a method to screen high numbers of sera rapidly for the presence of antibodies when the appropriate synthetic peptides are used as antigens.

4. CONCLUSIONS

The analysis of humoral immune responses using synthetic peptide antigens to detect specific antibodies in patients is extremely useful. The analysis allows early and rapid development of specific and sensitive test systems when a new infectious agent has been isolated and characterized. It is also very useful for distinguishing between similar subtypes of viruses or other infectious agents. In this regard, the use of synthetic peptides has become as important as nucleic acid technology for detecting pathogenic microorganisms.

ACKNOWLEDGMENTS. The authors thank Dr. Wolfgang Jilg for many helpful discussions and Dr. G. Unger and Dr. R. Biswas for information concerning diagnostic test systems containing synthetic peptide antigens which are licensed in Germany and the United States.

REFERENCES

1. Merrifield, R. B., 1963, Peptides synthesis. I. The synthesis of a tetrapeptide, *J. Am. Chem. Soc.* **85:**2149–2154.
2. Atherton, E., Fox, H., Logan, C. J., Harkiss, D., Sheppard, R. C., and Williams, B. J., 1978, A mild procedure for solid phase peptide synthesis: Use of fluorenylmethyloxycarbonyl amino acids, *J. Chem. Commun.* **13:**537–539.

3. Geysen, H. M., Meloen, R. H., and Barteling, S. J., 1984, Use of a peptide synthesis to probe viral antigens for epitopes to a resolution of a single amino acid, *Proc. Natl. Acad. Sci.* USA **81:**3998–4002.
4. Geysen, H. M., 1985, Antigen–antibody interactions at the molecular level: Adventures in peptide synthesis, *Immunol. Today* **6:**364–378.
5. Houghten, R. A., 1985, General method for rapid solid-phase synthesis of large numbers of peptides: Specificity of antigen-antibody interaction at the level of individual amino acids, *Proc. Natl. Acad. Sci.* USA **82:**5131–5135.
6. Jung, G. and Beck-Sickinger, A., 1992, Multiple peptide synthesis methods and their application, *Angew. Chemie.* **31:**367–372.
7. Neurath, A. R., Kent, S. H. B., and Strick, N., 1982, Specificity of antibodies elicited by a synthetic peptide having a sequence in common with a fragment of a virus protein, the hepatitis B surface antigen, *Prox. Natl. Acad. Sci. USA* **79:**7871–7875.
8. Kotwal, G. J., Baroudy, B. M., Kuramoto, I. K., McDonald, F. F., Schiff, G. M., Holland, P. V., and Zeldis, J. B., 1992, Detection of acute hepatitis C virus infection by ELISA using a synthetic peptide comprising a structural epitope, *Proc. Natl. Acad. Sci. USA* **89:**4486–4489.
9. Burns, J. M., Jr., Shreffler, W. G., Rosman, D. E., Sleath, P. R., March, C. J., and Reed, S. G., 1992, Identification and synthesis of a major conserved antigenic epitope of *Trypanosoma cruzi*, *Proc. Natl. Acad. Sci. USA* **89:**1239–1243.
10. Fargeas, C., Hommel, M., Maingon, R., Dourado, C., Monsigny, M., and Mayer, R., 1996, Synthetic peptide based enzyme-linked immunosorbant assay for serodiagnosis on visceral leishmaniasis, *J. Clin. Microbiol.* **34:**241–248.
11. Gnann, J. W. Jr, McCormick, J. B., Mitchell, S., Nelson, J. A., and Oldstone, M. B. A., 1987, Synthetic peptide immunoassay distinguishes HIV type 1 and HIV type 2 infections, *Science* **237:**1346–1349.
12. Roberts, C. R., Mitra, R., Hyams, K., Brodine, S. K., and Lal, R. B., 1992, Serologic differentiation of human T lymphotropic viruses type I from type II infection by synthetic peptide assays, *J. Med. Virol.* **36:**298–302.
13. Langedijk, J. P., Middel, W. G., Schaaper, W. M., Meloen, R. H., Kramps, J. A., Brandenburg, A. H., and van Oirschot, J. T., 1996, Type-specific diagnosis of respiratory syncytial virus infection, based on a synthetic peptide of the attachment protein G, *J. Immunol Methods* **193:**157–166.
14. Rapp, W., Zhang, L., Häbich, R., and Bayer, E., 1988, in: *Peptides 1988* (G. Jung and E. Bayer, eds.), Walter de Gruyter, Berlin, pp. 199–201.
15. Frank, R. and Döring, R., 1988, Simultaneous multiple peptide synthesis under continous flow conditions on cellulose filter discs as segmental solid supports, *Tetrahydron* **44:**6031–6035.
16. Fodor, S. P. A., Read, J. L., Pirrung, M. C., Struyer, L., Lu, A. T., and Solas, D., 1991, Light-directed spatially addressable parallel chemical synthesis, *Science* **251:**767–769.
17. Geyson, H. M., Rodda, S. J., Mason, T. J., Tribbick, G., and Schoofs, P. G., 1987, Strategies for epitope analysis using peptide synthesis, *J. Immunol. Methods* **102:**259–274.
18. van der Zee, R., van Eden, W., Meloen, R. H., Noordzij, A., and van Embeden, J. D., 1989, Efficient mapping and characterization of a T-cell epitope by simultaneous synthesis of multiple peptides, *Eur, J. Immunol.* **19:**43–49.
19. Krchnàk, V., Vaǵner, J., Novak, J., Suchankova, A., and Roubal, J., 1990, A general procedure for evaluation of immunological relevance of synthetic peptides: Peptides synthesized on paper in enzyme-linked immunosorbant assay, *Anal. Biochem.* **189:**80–83.
20. Birnbaum, S. and Mosback, K., 1992, Peptide screening, *Curr. Opinion Biotechnol.* **3:**49–54.
21. Houghten, R. A., Pinella, C., Blondelle, S. E., Appel, J. R., Dooley, C. T., and Cuero, J. H., 1991, Generation and use of synthetic peptide combinatoral libraries for basic research and drug discovery, *Nature* **354:**84–87.

22. Lam, K. S., Salmon, S. E., Hersh, E. M., Hruby, V. J., Kazmierski, W. M., and Knapp, R. J., 1991, A new type of synthetic peptide library for identifying ligand-binding activity, *Nature* **354:**82–84.
23. Pinilla, C., Appel, J. R., and Houghten, R. A., 1995, Synthetic peptide libraries for research and drug discovery, in: *Immunological Recognition of Peptides in Medicine and Biology* (N. D. Zergers, W. J. A. Boersma, and E. Claassen, eds.), CRC Press, Boca Ranton, New York, London, Tokyo, pp. 1–14.
24. Pinilla, C., Appel, J. R., Blanc, P., Houghten, R. A., 1992, Rapid identification of high affinity peptide ligands using positional scanning synthetic peptide combinatorial libraries, *BioTechniques* **13:**901–905.
25. Ostresh, J. M., Winkle, J. H., Hamashin, V. T., and Houghten, R. A., 1994, Peptide libraries: Determination of relative reaction rates of protected amino acids in competitive couplings, *Biopolymers* **34:**1681–1689.
26. Dooley, C. T. and Houghten, R. A., 1993, The use of positional scanning synthetic peptide combinatoral libraries for the rapid identification of opiod receptor ligands, *Life Sci.* **52:**1509–1516.
27. Smith, G. P. and Scott, J. K., 1993, Libraries of peptides and proteins displayed on filamentous phage, *Methods Enzymol.* **217:**228–257.
28. Dybwad, A., Bogen, B., Natvig, J. B., Forre, O., Sioud, M., 1995, Peptide phage libraries can be an efficient tool for identifying antibody ligands for polyclonal sera, *Clin. Exp. Immunol.* **102:**438–442.
29. Reichlin, M., 1980, Use of glutaraldehyde as a coupling agent for proteins and peptides, *Methods Enzymol.* **70:**159–165.
30. Arnon, R., Sela, M., Parant, M., and Chedid, L., Antiviral response elicited by a completely synthetic antigen with a built-in adjuvanticity, 1980, *Proc. Natl. Acad. Sci. USA* **77:**6769–6772.
31. Liu, F. T., Zinnecker, M., Hamaloda, T., and Katz, D. H., 1979, New procedure for preparation and isolation of conjugates of proteins and a synthetic copolymer of D-amino acids and immunochemical characterization of such antigens, *Biochemistry* **18:**609–687.
32. Modrow, S., Höflacher, B., Mertz, R., and Wolf, H., 1989, Production of carrier-bound synthetic peptides: Use as anitgens in HIV-1 ELISA-tests and for the elicitation of anitbodies, *J. Immunol. Methods.* **118:**1–7.
33. Modrow, S., Höflacher, B., Gürtler, L., Deinhardt, F., and Wolf, H., 1989, Carrier-bound synthetic oligopeptides in ELISA test systems for distinction between HIV-1 and HIV-2 infection, *J. Aquired Immune Defic. Sundr.* **2:**141–148.
34. Tang, X. L., Tregear, G. W., White, D. O., and Jackson, D. C., 1988, Minimun requirements for immunogenic and antigenic activities of homologs of a synthetic peptide of influenza hemagglutinin, *J. Virol.* **62:**4745–4751.
35. Meloen, R. H., and Barteling, S. J., 1986, Epitope mapping of the outer structural protein VP1 of three different serotypes of FMDV, *Virology* **149:**55–63.
36. Langeveld, J. P. M., Casal, J. I., Vela, C., Dalsgaard, K., Smale, S. H., Puijk, W. C., and Meloen, R. H., 1993, B-cell epitopes of canine parvovirus: Distrubution on the primary structure and exposure at the viral surface, *J. Virol.* **68:**765–772.
37. Clarke, J. L., Drake, A. F., Wallace, G. R., Allen, A. K., Kelly, J. M., and Miles, M. A., 1990, PEPSCAN and circular dichroism analysis of a repetitive antigen of *Trypanosoma curzi, Biochem. Soc. Trans.* **18:**868–869.
38. Miles, M. A., Wallace, G. R., and Clarke, J. L., 1989, Multiple peptide synthesis (PEPSCAN method) for the systematic analysis of B- and T-cell epitopes: Application to parasite proteins, *Parasitology Today* **5:**397–406.

39. Appel, J. R., Pinilla, C., and Houghten, R. A., 1992, Identification of related peptides recognized by a monoclonal antibody using a peptide combinatoral library, *Immunomethods* **1:**17–24.
40. Modrow, S. and Wolf, H., 1990, Use of synthetic peptides as diagnostic reagents in virology, in: *Immunochemistry of Viruses,* Volume II (M. H. V. van Regenmortel and A. R. Neurath, eds.), Elsevier, Amsterdam, New York, Oxford, pp. 83–102.
41. Schwarz, T. F., Modrow, S., Hottenträger, B., Höflacher, B., Jäger, G., Schartl, W., Sumazaki, R., Wolf, H., Middledorp, J., Roggendorf, M., and Deinhardt, F., 1991, New oligopeptide IgG test for human parvovirus B19 antibodies, *J. Clin. Microbiol.* **29:**431–435.
42. Rötzschke, O., Falk, K., Deres, K., Schild, H., Norda, M., Metzger, J., Jung, G., and Rammensee, H. G., 1990, Isoloation and analysis of naturally processed viral peptides as recognized by cytotoxic T-cells, *Nature* **348:**252–254.
43. Brown, J. H., Jardetzky, T. S., Gorga, J. C., Stern, L. J., Urban, R. G., Strominger, J. L., and Wiley, D. C., 1993, Three-dimensional structure of human class II histocompatibility antigen DDR1, *Nature* **364:**33–39.
44. Bogedain, C., Modrow, S., Wolf, H., and Jilg, W., 1995, Specific cytotoxic T-lymphocytes recognize the immediate-early transactivator Zta of Epstein–Barr virus, *J. Virol.* **69:**4872–4879.
45. van Eden, W., Noordzij, A., and Hensen, E. J., 1989, A modified PEPSCAN method for rapid identification and characterization of T-cell epitopes in protein antigens, in: *Vaccines 89,* Cold Spring Harbor Laboratory Press; Cold Spring Harbor, N.Y., pp. 33–39.
46. Stuber, G., Modrow, S., Höglund, P., Wolf, H., and Klein, G., 1992, Assessment of major histocompatibility complex class I interaction with Epstein–Barr virus and human immunodeficiency virus peptides by elevation of H-2 and HLA in peptide loading deficient cells, *Eur. J. Immunol.* **22:**2697–2703.
47. Hopp, T. P. and Woods, K. R., 1981, Prediction of protein antigenic determinants from amino acid sequences, *Proc. Natl. Acad. Sci. USA* **78:**3824–3827.
48. Kyte, J. and Doolittle, R. F., 1987, A simple method for displaying the hydrophobic character of a protein, *J. Mol. Biol.* **157:**105–132.
49. Westhof, E., Altschuh, D., Moras, D., Bloomer, D., Mondragon, A., Klug, A., and van Regenmortel, M. H., 1985, Correlation between segmental mobility and the location of antigenic deerminants in proteins, *Nature* **311:**123–125.
50. Tainer, J. A., Getzhof, E. D., Patterson, Y., Olson, A. J., and Lerner, R. A., 1985, The atomic mobility components of protein antigenicity, *Ann. Rev. Immunol.* **3:**501–535.
51. Karplus, P. A. and Schulz, G. E., 1985, Prediction of chain flexibility in proteins, *Natruwissenschaften* **72:**212–214.
52. Janin J. and Wodak, S., 1978, Conformation of amino acid side chains in proteins, *J. Mol. Biol.* **125:**357–386.
53. Emini, E. A., Hughes, J. V., Perlow, D. S., and Boeger, J., 1985, Induction of hepatitis-A-virus neutralizing antibody by a virus-speicfic peptide, *J. Virol.* **55:**836–839.
54. Bittle, J. L., Houghten, R. A., Alexander, H., Shinnick, T. M., Sutcliffe, J. G., Lerner, R. A., Rowlands, D. J., and Brown, F., 1982, Protection against foot-and-mouth-disease by immunization with a chemically synthesized peptide predicted from the viral nucleotide sequence, *Nature,* **298:**30–33.
55. Modrow, S., Hahn, B. H., Shaw, G. M., Gallo, R. C., Wong-Stall, F., and Wolf, H., 1987, Computer-assisted analysis of the envelopoe protein sequences of seven HTLV-III/LAV isolates: Prediction of antigenic epitopes in conserved and variable regions, *J. Virol.* **61:**570–578.
56. Chou, P. Y. and Fasman, G. D., 1978, Prediction of the secondary structure of proteins from their amino acid sequence, *Adv. Enzymol.* **47:**45–148.
57. Garnier, J., Osguthrope, O. J., and Robson, B., 1978, Analysis of accuracy and implication of simple methods for predicting the secondary structure of globular proteins, *J. Mol. Biol.* **120:**97–120.

58. Wolf, H., Modrow, S., Motz, M., Jameson B., Hermann, G., and Foŕtsch, B., 1988, An integrated family of amino acid sequence analysis programs, *CABIOS* **4:**187–191.
59. Modrow, S. and Wolf, H., 1995, Progress and frontiers in prediction of B- and T-cell epitopes, in: *Immunological Recognition of Peptides in Medicine and Biology* (N. D. Zergers, W. J. A. Boersma, and E. Claassen, eds.), CRC Press, Boca Raton, New York, London, Tokyo, pp. 125–131.
60. Björkman, P. J., Saper, M. A., Samraoui, B., Bennett, W. S., Strominger, J. L., and Wiley, D. C., 1987, Structure of the human class I histocompatibility antigen HLA-A2, *Nature* **329:**506–510.
61. Björkman, P. J. and Parham, P., 1990, Structure, function and diversity of class I major histocompatibility complex molecules, *Ann. Rev. Biochem.* **59:**253–279.
62. Saper, M. A., Bjorkman, P. J., and Wiley, D. C., 1991, Refined structures of the human histocompatibility antigen HLA-A2 at 2.6 Å resolution, *J. Mol. Biol.* **219:**277–295.
63. Falk, K. and Rötzschke, O., 1993, Consensus motifs and peptide ligands of MHC class I molecules, *Sem. Immunol.* **5:**81–89.
64. Buus, S., Pederson, L. O., and Stryhn, A., 1995, The analysis of peptide binding to MHC and of MHC specificity, in: *Immunological Recognition of Peptides in Medicine and Biology* (N. D. Zergers, W. J. A. Boersman, and E. Claassen, eds.), CRC Press, Boca Raton, New York, London, Tokyo, pp. 61–77.
65. Lackmann, M., Rajasekariah, P., Iismaa, S. E., Jones, G., Cornish, D. J., Hu, S., Simpson, R. J., Moritz, R. L., and Geczy, C. R., 1993, Identificaiton of a chemotactic domain of the proinflammatory S100 protein SP-10, *J. Immunol.* **150:**2981–2991.
66. Niedrig, M., Gelderblom, H., Pauli, G., März, J., Bikhard, H., Wolf, H., and Modrow, S., 1994 Inhibition of infectious HIV-1 particle formation by Gag-protein derived peptides, *J. Gen. Virol.* **75:**1469–1474.
67. Perira, H. A., Erdem, I., Pohl, J., and Spitznagel, J. K., 1993, Synthetic bactericidal peptide based on CAP37: A 37 kDa human neutrophil granule-associated cationic antimicrobial protein chemotactic for monocytes, *Proc. Natl. Acad. Sci. USA* **90:**4733–4737.
68. Telford, E., Lankenen, H., and Marsden, H., 1990, Inhibition of equine herpesvirus type 1-induced ribonucleotide reductase by a nonapeptide YAGAVVNDL, *J. Gen. Virol.* **71:**1373–1378.
69. Rosa, C., Osborne, S., Garetto, F., Griva, S., Rivella, A., Calabresi, G., Guaschino, R., and Bonelli, F., 1995, Epitope mapping the the NS4 and NS5 gene products of hepatitis C virus and the use for a chimeric NS4-NS5 peptide for serodiagnosis, *J. Virol. Methods* **55:**219–232.
70. Tosswill, J. H., McAlpine, L., and Mortimer, P. P., 1993, Serological specifiction of human T-cell leukemia virus infections using synthetic peptide antigens, *J. Med. Virol.* **40:**83–85.
71. Brattegaard, K., Soroh, D., Zadi, F., Digbeu, H., Vetter, K. M., and De Cock, K. M., Insensitivity of a synthetic peptide based test (Pepti-LAV 1-2) for the diagnosis of HIV-infection in African children, *AIDS* **9:**656–657.
72. Chamekh, M., Gras-Masse, H., Bossus, M., Facon, B., Dissous, C., Tartar, A., and Capron, A., 1992, Diagnostic value of a synthetic peptide derived from *Echinococcus granulosus* recombinant protein, *J. Clin. Invest.* **89:**458–464.
73. Ferrerira-Junior, O. C., Vaz, R. S., Carvalho, M. B., Guerra, C., Fabron, A. L., Rosemblit, J., and Hamerschlak, N., 1995, Human T-lymphotropic type I and type II infections and correlation with risk factors in blood donors from Sao Paulo, Brazil, *Transfusion* **35:**258–263.
74. Brattegaard, K., Kouadio, J., Adom, M. L., Doorly, R., George, J. R., and De Cock, K. M., 1993, Rapid and simple screening and supplemental testing for HIV-1 and HIV-2 infections in West Africia, *AIDS* **7:**883–885.
75. Todd, J., Kink, J., Leahy, D., Preisel-Simmons, B., Laska, S., Wolff, P., Byrne, R., Babler, S., and

McCoy-Haman, M., 1992, A novel semi-automated paramagnetic microparticle based enzyme immunoassay for hepatiatis C virus; its application to serologic testing, *J. Immunology* **13:**393–410.

76. Wilson, K. M., Gerometta, M., Rylatt, D. B., Bundesen, P. G., McPhee, D. A., Hillyard, C. J., and Kemp, B. E., 1991, Rapid whole blood assay for HIV-1 seropositivity using an Fab-peptide conjugate, *J. Immunol Methods* **138:**111–119.

3

Flow Cytometric Analysis of Virally Infected Cells

In Vitro and *in Vivo* Studies

JAMES J. McSHARRY

1. INTRODUCTION

Virus–cell interactions have been analyzed by using fluorochrome-labeled antibodies in conjunction with flow cytometry for more than 20 years. *In vitro* studies of virus–cell interactions include the detection and quantitation of (1) the binding of fluorochrome-labeled viruses to their receptors on cell surfaces; (2) viral antigens on the cell surface, in the cytoplasm, and in the nucleus of virus infected cells; (3) virally induced apoptosis; and (4) the effects of viral infection on the expression of cellular antigens and nucleic acid synthesis. Some investigators have used fluorochrome-labeled monoclonal antibodies and flow cytometry to detect and quantitate virally infected cells directly in clinical specimens.[1–6] The published literature has recently been reviewed.[7] In this chapter, I introduce the principles of flow cytometry, briefly review some of the published reports on the use of fluorochrome labeled antibodies and flow cytometry for directly detecting viruses in clinical specimens, and present some recent findings on the use of fluorochrome-labeled antibodies and flow cytometry for determining the effect of

JAMES J. McSHARRY • Department of Microbiology, Immunology, and Molecular Genetics, Albany Medical College, Albany, New York 12208.

Rapid Detection of Infectious Agents, edited by Specter *et al.* Plenum Press, New York, 1998.

antiviral drugs on viral antigen synthesis in human cytomegalovirus (HCMV)-infected cells and the drug sensitivity of HCMV clinical isolates.

2. A BRIEF DESCRIPTION OF FLOW CYTOMETRY

Flow cytometry is defined as the measurement of the physical and/or chemical characteristics of cells while they pass single file in a fluid stream through a measuring apparatus.[8] Simple flow cytometers use a single argon ion laser to simultaneously measure a number of cellular properties, including cell number, cell size, cell granularity, and cell-associated fluorescence. The light-scattering properties of each cell are measured by collecting scattered laser light with lenses set to collect forward angle (FW-SC) and right-angle (RT-SC) light scatter. This information is used to determine the number, size, and granularity of analyzed cells in the sample. If the cells passing through the laser beam are labeled with one or more fluorochromes directed against specific cellular and/or viral components, the laser light excites the fluorochrome(s) causing each fluorochrome to emit light at a higher wave length. The emitted light is separated into individual components by lenses, captured by photomultiplier tubes, digitized, and the data are displayed on a monitor screen, printed on paper to yield a hard copy of the data, and stored in a computer as a permanent file that can be used for further data analysis at a future time. Thus, the flow cytometric analysis of fluorochrome-labeled cells yields information on the number, size, and granularity of the cells and the amount of fluorochrome associated with each cell. The analysis of the light-scattering properties of cells is often used to separate different cell populations in a sample on the basis of cell size and granularity. Fluorochrome-labeled antigens and nucleic acids are used to further identify cells within each of the separated populations. Several thousand cells are analyzed per second yielding statistically significant data in a short time. In this manner, the physical and chemical properties of each cell passing through the laser beam are examined.

Figure 1 is an illustration of a generic flow cytometer that uses a single argon ion laser to distinguish the light-scattering properties of cells and three different fluorochromes associated with the cells. More sophisticated flow cytometers are also available that use two or more lasers to examine additional fluorochromes. Some flow cytometers are cell sorters that physically separate and collect specific cells in a population. However, the experiments described in this chapter require only an instrument with a single argon ion laser that analyzes the light-scattering properties of cells and two or three cell-associated fluorochromes simultaneously.

A typical analysis of uninfected and HCMV-infected fibroblasts treated with 7-amino actinomycin D (7-AAD), fluorescein isothiocyanate (FITC)-labeled

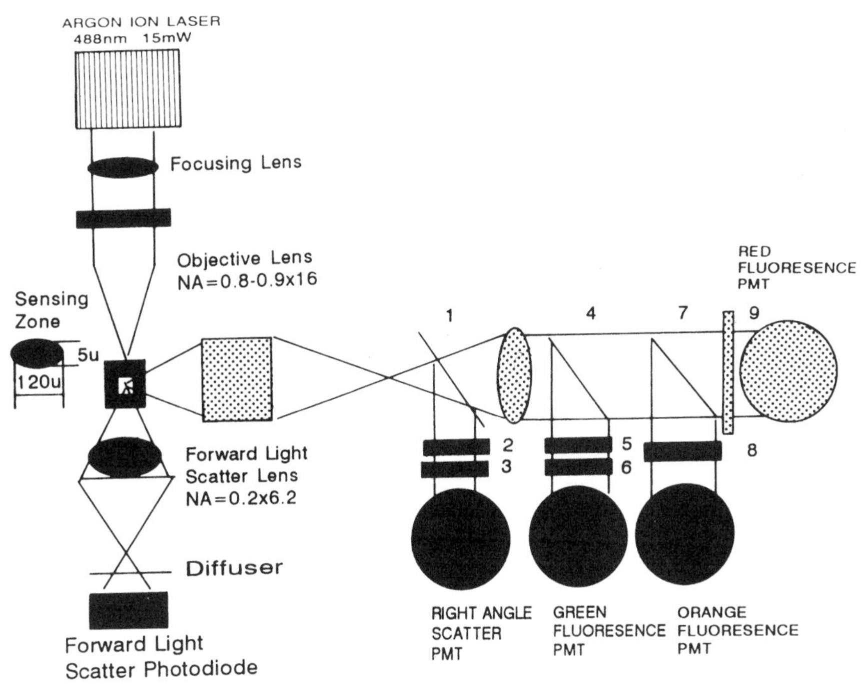

FIGURE 1. A flow cell with attached optical systems. As the cells pass through the flow cell (the small, black box in the center left), a laser beam intersects the stream of cells and scatters light. Forward angle light scatter passes through the forward light-scattering lens, and the energy is collected by the forward light-scattering photodiode. Right-angle light scatter passes through the objective lens, the beam splitter, the laser line filter, and the diffuser, and the energy is collected by the right-angle scatter photomultiplier tube (PMT). If the cells are labeled with fluorescent molecules, the laser excites these molecules, which emit light of higher energies. The emitted energies of different wavelengths pass through the objective lens and various filters and are collected and amplified by the various PMTs. The amplified signals are converted into digital information and stored in a computer for further analysis. Numbers 1 through 9 refer to the following: 1, beam splitter; 2, laser line filter, 396–496 nm band-pass; 3, diffuser; 4, dichroic mirror 1, 570-nm long-pass filter; 5, laser-cut filter, 490-nm short cut; 6, green filter, 515–530-nm band-pass; 7, dichroic mirror 2, 610-nm long-pass; 8, orange filter, 565–592-nm band-pass; and 9, red filter, 660-nm long-pass (from Ref. 7 with permission).

monoclonal antibody to an HCMV immediate-early (IE) antigen, and phycoerythrin (PE)-labeled monoclonal antibody to an HCMV late antigen is shown in Fig. 2. The left-hand set of panels are dot blots of the analysis of uninfected cells treated with fluorochrome-labeled antigen-specific monoclonal antibodies. When uninfected cells are analyzed for RT-SC versus FW-SC, the light scatter pattern

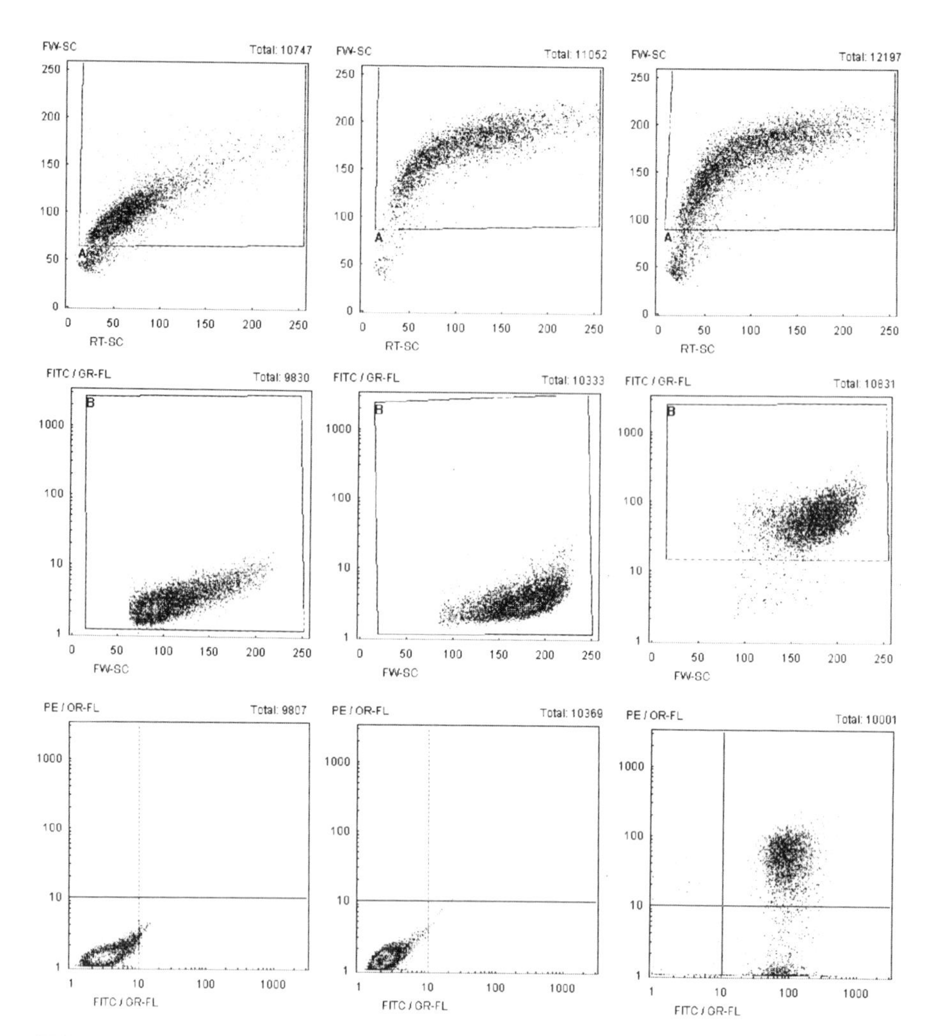

FIGURE 2. Triple-color, flow cytometric analysis of MRC-5 cells. Uninfected or HCMV-infected cells were permeabilized with 90% methanol, treated with FITC-labeled monoclonal antibody to the HCMV IE antigen, PE-labeled monoclonal antibody to an HCMV late antigen, or FITC- and PE-labeled isotype control monoclonal antibodies, 7-AAD, and analyzed for three-color fluorescence by flow cytometry. The left-hand column shows dot blots of infected cells treated with fluorochrome-labeled monoclonal antibodies to HCMV IE and late antigens and 7-AAD; the middle column shows dot blots of HCMV-infected cells treated with FITC- and PE-labeled isotype control monoclonal antibodies and 7-AAD; the right-hand column shows dot blots of HCMV-infected cells treated with FITC- and PE-labeled monoclonal antibodies to HCMV IE and late antigens and 7-AAD. See text for details of analysis.

distribution is characteristic of uninfected intact cells. Gate A separates the intact cells from the debris. Analysis of the gated uninfected cells for fluorescein isothiocyanate/green fluorescence (FITC/GR-FL) versus FW-SC results in a fairly homogenous population of cells with low FITC fluorescent intensity (middle graph). When the uninfected cells are analyzed for phycoerythrin/orange fluorescence (PE/OR-FL) versus FITC/GR-FL intensities, essentially all of the events fall into the lower left-hand quadrant which indicates that they have low fluorescent intensities and are negative for the specific viral antigens (bottom graph). The middle panels of dot blots are HCMV-infected cells treated with FITC- and PE-labeled, isotype control, monoclonal antibodies. When these HCMV-infected cells are analyzed for RT-SC versus FW-SC, the majority of cells exhibit RT-SC intensity and FW-SC properties that are greater than those exhibited by the uninfected cells (upper graph). The increased RT-SC and FW-SC are caused by the increased DNA content and cell size associated with HCMV-infected cells. Gate A separates the cells from the debris. Analysis of the intact cells in gate A for FITC/GR-FL intensity versus FW-SC shows that the majority of the cells have low FITC/GR-FL intensity and that they are larger than the uninfected cells (middle graph). When the HCMV-infected cells treated with isotype control monoclonal antibodies are analyzed for PE/OR-FL versus FITC/GR-FL intensities, all of the cells exhibit low FITC/GR-FL and PE/OR-FL intensities (lower graph). The right-hand panel of dot blots represents the analysis of HCMV-infected cells treated with fluorochrome-labeled, antigen-specific, monoclonal antibodies. When the HCMV-infected cells are analyzed for RT-SC versus FW-SC, the majority of cells exhibit RT-SC are larger than the uninfected cells (upper graph). The increased RT-SC and FW-SC are caused by the increased DNA content and cell size of HCMV-infected cells. Gate A separates cells from debris. Analysis of the cells in gate A for FITC/GR-FL intensity versus FW-SC shows that the majority of HCMV-infected cells have increased FITC/GR-FL intensity and are larger than the uninfected cells (middle graph). When the FITC-labeled cells in gate B are analyzed for FITC/GR-FL intensity versus PE/OR-FL intensity, all of the cells exhibit increased FITC/GR-FL intensity, and a portion of the cells exhibit both FITC/GR-FL and PE/OR-FL intensities (lower graph). By using the gates set on the uninfected cells or the HCMV-infected cells treated with isotype control antibodies, it is possible to determine the percentage of HCMV-infected cells that are expressing each antigen detected by the antigen-specific, fluorochrome-labeled monoclonal antibody. Analysis of the HCMV-infected cells for PE/OR-FL versus FITC/GR-FL intensities shows that 100% of the cells express the IE antigen and 34.3% of the cells express both the IE and the late antigen. Using fluorochrome-labeled antibodies to any viral antigen and fluorochrome-labeled nucleic acid dyes for detecting intact cells, this flow cytometry assay is used to detect and quantitate any virally infected cell.

3. DETECTION OF VIRALLY INFECTED CELLS IN CLINICAL SPECIMENS

3.1. Human Immunodeficiency Virus (HIV)

A number of groups have demonstrated that fluorochrome-labeled monoclonal antibodies in conjunction with flow cytometry directly detect and quantitate HIV-infected cells in peripheral blood. In each case, peripheral blood was obtained from HIV-seronegative controls, HIV-seropositive asymptomatic patients, and patients with AIDS. Peripheral blood mononuclear cells (PBMCs) were isolated on histopaque, permeabilized by several different methods, treated with fluorochrome-labeled monoclonal antibodies to HIV antigens, and analyzed by flow cytometry. McSharry *et al.*[2] permeabilized the PBMCs with 90% methanol, treated the cells with monoclonal antibodies to the HIV p24 or nef antigens followed by FITC-conjugated goat antimouse IgG $F(ab')_2$, RNAse, and propidium iodide (PI) to identify intact cells with a diploid or greater DNA content, and analyzed the cells for two-color fluorescence by flow cytometry. They showed that the percentage of HIV-infected PBMCs increases with the severity of disease and is inversely proportional to the number of $CD4^+$ lymphocytes. This group was the first to show that AIDS patients have a large number of HIV-infected PBMCs in their circulation. Ohlssom-Wilhelm *et al.*[3] independently showed that fluorochrome-labeled monoclonal antibodies to the HIV p24 antigen and flow cytometry detect and quantitate HIV-infected cells in PBMCs. By excluding the monocyte population from their analysis, they showed that the virally infected cells are $CD4^+$ lymphocytes. Both groups used this technology to show that AZT treatment reduces the percentage of HIV-infected PBMCs in the circulation.[2,3] These results have been reproduced by others.[4–6] Although fluorochrome-labeled antibodies and flow cytometry detect and quantitate HIV-infected cells in clinical specimens and monitor the effect of antiretroviral therapy in patients with AIDS, under the experimental conditions used this technique fails to detect virally infected cells in asymptomatic HIV-seropositive patients.

3.2. Human Cytomegalovirus (HCMV)

Fluorochrome-labeled HCMV-infected cells in bronchoalveolar lavage (BAL) specimens and PBMCs have been detected and quantitated by flow cytometry. The initial studies used indirect immunofluorescence to label cells treated with a monoclonal antibody to an HCMV early antigen and flow cytometry to detect and quantitate HCMV-infected cells in BAL specimens obtained from patients with confirmed HCMV pneumonia.[1] In comparison with the standard tissue culture assay, the specificity and sensitivity of the flow cytometry assay approached 90%. Furthermore, the time required for a laboratory diagnosis was one day for the flow cytometry assay and four to six weeks for the standard tissue culture assay. More recently, fluorochrome-labeled monoclonal antibodies to the HCMV IE antigen and flow cytometry have been used to show that PBMCs obtained from AIDS patients or transplant patients with acute HCMV disease express the HCMV IE antigen.[7] This technology was used to monitor the effect of ganciclovir on the number of PBMCs expressing the HCMV IE antigen.[7] These preliminary studies showed that fluorochrome-labeled monoclonal antibodies in conjunction with flow cytometry detect and quantitate HCMV-infected cells directly in clinical specimens. The original studies used monoclonal antibodies to the HCMV IE antigen and analyzed lymphocytes and monocytes present in PBMCs. More recent studies suggest that the HCMV lower matrix protein pp65, present in the granulocyte population of peripheral blood, is a better target for detecting HCMV-infected cells in peripheral blood.[9–12] Further study targeting different white blood cell populations in the peripheral blood, including lymphocytes, monocytes, and granulocytes and using individual or combinations of monoclonal antibodies to IE, early (pp65), and late HCMV antigens in conjunction with flow cytometry is necessary to develop this technology for rapidly detecting HCMV-infected cells in clinical specimens.

3.3. Herpes Simplex Viruses

Herpes simplex viruses (HSV) replicate rapidly in tissue culture cells.[13] A laboratory diagnosis involving inoculation of a clinical specimen into tissue culture cells, production of a cytopathic effect (CPE), and confirmation with monoclonal antibodies is performed in a few days to a week. Fluorochrome-labeled monoclonal antibodies to HSV antigens and flow cytometry detect and quantitate HSV in clinical specimens after amplification in tissue culture.[14] Using monoclonal antibodies that recognize viral antigens shared by both HSV-1 and HSV-2, virally infected cells were detected and quantitated within one day of culture. The serotype of the clinical isolate was determined by treating the HSV-infected tissue culture cells with monoclonal antibodies to type specific antigens. To detect and quantitate HSV-infected tissue culture cells and to determine the HSV serotype simultaneously, HSV-infected cell cultures were treated with a cocktail consisting of FITC-labeled monoclonal antibody to an HSV type 1 specific antigen and a PE-labeled monoclonal antibody to an HSV type 2 specific antigen and analyzed by flow cytometry (McSharry, unpublished data). This rapid, quantitative, flow cytometric analysis has the advantage of automation and speed and is less labor-intensive and subjective than microscopic observation to determine the presence of CPE and fluorescence microscopy to confirm the laboratory diagnosis of HSV infection. Expanded use of fluorochrome-labeled monoclonal antibodies to viral antigens and flow cytometry for detecting and quantitating virally infected tissue culture cells will make the diagnostic laboratory more efficient and productive.

4. *IN VITRO* STUDIES WITH HUMAN CYTOMEGALOVIRUS

The use of indirect immunofluorescence and flow cytometry to detect and quantitate HCMV-infected tissue culture cells has been reported.[1,15] The results clearly show that uninfected and HCMV-infected cells treated with fluorochrome-labeled monoclonal antibodies and PI are detected and quantitated by flow cytometry and that uninfected cells are easily distinguished from HCMV-infected cells.

Using monoclonal antibodies to IE, early, and late HCMV antigens and an FITC-labeled second antibody and flow cytometry, the time course of the synthesis of these antigens in HCMV-infected cell cultures was determined.[15,16] The data show that IE antigens are synthesized within 2 hr postinfection and that their synthesis continues throughout the infection. Late antigen synthesis begins between 48 and 72 hr postinfection and reaches maximum production between 96 and 120 hr postinfection. This information has been used to develop procedures for determining the inhibitory dose 50 (IC_{50}) and (IC_{90}) for various antiviral compounds that inhibit HCMV synthesis.[17] Furthermore, this assay was developed into a rapid, quantitative method for determining the sensitivity of HCMV clinical isolates to ganciclovir.[18,19] During the remainder of this section, I describe the results of these experiments in some detail.

4.1. Materials and Methods

4.1.1. Cells and Medium

Monolayer cultures of MRC-5 cells (ATCC CCL 171) or human foreskin fibroblasts (HFF) were used throughout these experiments. Minimal essential medium (MEM), supplemented with 10% fetal bovine serum (FBS), penicillin, streptomycin, and amphotericin B, was used to grow the cells and the virus. MRC-5 cells at passage 20 to 25 and HFF at passage 9 to 15 were used for these experiments.

4.1.2. Viruses

The ganciclovir-sensitive AD169 strain of HCMV, its ganciclovir-resistant derivative, D6/3/1,[20] and HCMV clinical isolates were used. Viruses were grown in MRC-5 of HFF cell monolayers. Cell-free or cell-associated virus was inoculated onto cell monolayers at a multiplicity of infection (moi) of 0.1 to 10 in the presence of MEM supplemented with 10% FBS and antibiotics. After a 2-hr adsorption period at 37 °C, the inoculum was removed, MEM supplemented with 3% FBS and antibiotics was added, and the infected monolayers were incubated at 37 °C for various time periods.

4.1.3. Antiviral Compounds

Ganciclovir was a generous gift from Roche/Syntex Laboratories, Nutley, NJ; foscarnet was purchased from Sigma Chemical Co., St. Louis, MO; and cidofovir was a generous gift from Gilead Sciences, Foster City, CA.

4.1.4. Monoclonal Antibodies

Unconjugated and FITC- or PE-labeled monoclonal antibodies were generously provided by Chemicon International, Inc., Temecula, CA. FITC-conjugated goat antimouse IgG $F(ab')_2$ antibody was purchased from Boehringer Mannheim Co., Indianapolis, IN.

4.1.5. Preparation of Cells for Flow Cytometry

Uninfected and virally infected cells were removed from the culture flask with trypsin/EDTA, resuspended with MEM supplemented with 10% FBS, pelleted at 1000 RPM for 10 minutes, washed once with PBS without Ca^{2+} and MG^{2+}, and put on ice for 60 minutes. The cells were permeabilized by adding ice-cold absolute methanol to the residual PBS in the tube to yield a final concentration of 90% methanol. The permeabilized cells were stored at −70 °C until they were prepared for analysis by flow cytometry.

4.1.6. Treating Cells with Monoclonal Antibodies and DNA Binding Reagents

The permeabilized cells were incubated for 90 minutes at 37 °C with monoclonal antibodies diluted to the appropriate concentration in wash solution (50% goat serum in PBS, 0.002% Triton X-100, and 0.1% sodium azide). After the incubation period, the cells were washed three times with wash solution, treated with FITC-conjugated goat antimouse IgG $F(ab')_2$ antibody, and incubated at 37 °C for an additional 90 minutes. After the second incubation, the cells were washed as before, treated with RNAse (Sigma Chemical Company, St. Louis, MO) and PI (Sigma Chemical Company, St. Louis, MO), and analyzed for two-color fluorescence by flow cytometry. In some experiments, a combination of FITC- and PE-labeled monoclonal antibodies (Chemicon International, Inc.,

Temecula, CA) were used to label HCMV antigens, 7-AAD (Sigma Chemical Company,, St. Louis, MO) was used to stain the DNA, and the cells were analyzed for three-color fluorescence by flow cytometry.

4.1.7. Flow Cytometry

An Ortho Cytoronabsolute analytical flow cytometer (Ortho Diagnostic Systems, Raritan, NJ) was used for these studies. For two-color analysis, the cells were analyzed for DNA/RD-FL intensity along the Y-axis and FW-SC along the X-axis. Cells with a diploid or greater DNA content were gated and analyzed for FITC-fluorescence intensity (FITC/GR-FL) along the Y-axis and PI-fluorescence intensity (DNA/RD-FL) along the X-axis. Gates were set on the uninfected cells so that less than 1% of the uninfected cells were in the gated area. Using these gates, virally infected cells exhibited FITC/GR-FL intensity at least tenfold greater than that exhibited by the uninfected cells. A similar set of gates was set for three-color analysis. Initially, the cells were analyzed for red (7-AAD)-fluorescent intensity versus FW-SC. Cells with a diploid or greater DNA content were gated and analyzed for FITC/GR-FL intensity versus FW-SC. Virally infected cells exhibiting FITC/GR-FL intensity above background were gated and analyzed for PE/OR-FL-fluorescent intensity versus FITC/GR-FL intensity. The percentages of cells in positive gates were calculated, printed, and stored in the computer for further data analysis.

4.2. Assay for the IC_{50} and IC_{90} of Antiviral Drugs for Human Cytomegalovirus

Currently licensed antiviral compounds in clinical use for treating acute HCMV infections, such as ganciclovir, cidofovir, and foscarnet, inhibit the synthesis of HCMV DNA.[21] During the replication of HCMV, IE and early antigens are synthesized from messenger RNAs transcribed from the parental genome, followed by DNA synthesis, and the synthesis of late antigens from messenger RNAs transcribed from progeny DNA genomes.[22] Because HCMV IE antigen synthesis is independent of viral DNA synthesis and HCMV late-antigen synthesis depends on viral DNA synthesis, antiviral agents that block HCMV DNA synthesis should inhibit the synthesis of late antigens in HCMV-infected cells, but not IE antigen synthesis. Flow cytometry has been used to detect and quantitate cells synthesizing the HCMV IE and late antigens.[16] This procedure should be able to quantitate the effect of antiviral drugs that block viral DNA synthesis or the synthesis of late antigens in HCMV-infected cells. Furthermore, by using various concentrations of drugs that block viral DNA synthesis, the inhibitory concentrations of these drugs for the infecting virus could be determined. To test this hypothesis, MRC-5 cells were infected with the drug-sensitive AD169 strain of

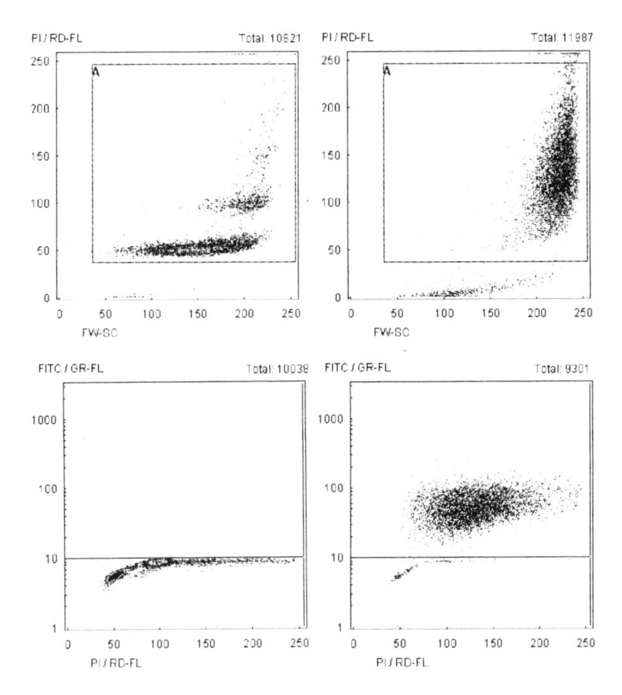

FIGURE 3. Two-color, flow cytometric analysis of MRC-5 cells. Separate samples of uninfected and HCMV-infected MRC-5 cells were harvested, permeabilized with 90% methanol, and treated with either a monoclonal antibody to an HCMV IE antigen or a monoclonal antibody to an HCMV late antigen, followed by FITC-conjugated second antibody, RNAse and PI, and analyzed for two-color fluorescence by flow cytometry. The upper panels are analyses of DNA content (PI/RD-FL) versus cell size (FW-SC). The lower panels are analyses of the gated cells from the respective upper panels for IE or late viral antigen (FITC/GR-FL) versus DNA content (PI/RD-FL). See text for details.

HCMV at a multiplicity of infection of 10 and virus was allowed to grow in the absence of presence of various concentrations of antiviral drugs for 96 h. The cells were harvested, treated with monoclonal antibodies to IE or late antigens, FITC-second antibody, RNAse and PI, and analyzed for two color fluorescence by flow cytometry. Figure 3 shows a typical analysis of two-color fluorescence of uninfected and HCMV-infected cells in the absence of antiviral drugs. Uninfected intact cells have a PI binding pattern characteristic of cells with a diploid or greater DNA content illustrated by the major band at 50 units on the Y-axis and minor bands at 100 units and higher. Gate A was set around all of the cells that bind PI, and the intact cells were analyzed for FITC/GR-FL intensity versus PI/RD-FL intensity. Uninfected cells have essentially no FITC/GR-FL intensity above the gate set at the first decade on the Y-axis. Analysis of HCMV-infected cells for

TABLE I
Flow Cytometric Determination of IC_{50} and IC_{90} of Ganciclovir, Cidofovir, and Foscarnet on Drug-Sensitive AD169-Infected MRC-5 Cells

	μM Ganciclovir			
	0	1.5	3.0	12.0
	% of cells expressing the HCMV antigens			
Isotype antibody	1.2	1.0	1.3	0.9
HCMV IE antibody	97.2	96.3	98.0	98.1
HCMV late antibody	88.1	45.7	30.5	6.0

	μM Cidofovir				
	0	0.6	1.25	2.5	5.0
	% of cells expressing the HCMV antigens				
Isotype	1.3	1.9	2.0	1.9	2.8
HCMV IE antibody	74.7	78.7	84.1	84.03	80.3
HCMV late antibody	46.2	30.0	18.3	6.7	2.0

	μM Foscarnet				
	0	100	200	400	800
	% of cells expressing the HCMV antigens				
Isotype	0.5	1.8	2.9	1.9	2.3
HCMV IE antibody	69.7	66.4	68.3	69.7	60.6
HCMV late antibody	43.4	25.7	20.6	14.7	10.0

PI/RD-FL intensity versus FW-SC shows that the infected cells bind more PI and are much larger than the uninfected cells. When the HCMV-infected cells in gate A are analyzed for FITC/GR-FL intensity versus PI/RD-FL intensity, a large proportion of the infected cells exhibit FITC/GR-FL intensity above the gate set at the first decade on the Y-axis. The data in Table 1 show the effect of various concentrations of ganciclovir, codifovir, and foscarnet on the synthesis of IE and late antigens in cells infected with the drug-sensitive AD169 strain of HCMV. In the absence of ganciclovir, 97.2% of the virally infected cells express the IE antigen, and 88.1% of the cells express the late antigen. In the presence of 1.5 μM ganciclovir, 96.3% of the cells express the IE antigen, and 45.7% of the cells express the late antigen. At 12 μM ganciclovir only 6% of the cells express the late antigen. These results suggest that the IC_{50} and IC_{90} for this strain of virus for ganciclovir are approximately 1.5 μM and 12 μM, respectively. Similar results are shown for foscarnet and cidofovir. These results suggest that this technology can be used to determine the IC_{50} and IC_{90} for various antiviral drugs active against HCMV. A similar approach has been used to determine the IC_{50} of antiviral compounds that inhibit viral DNA synthesis in HSV-infected cells.[23] This rapid, quantitative technology could be used to determine the mode of action of antiviral compounds that inhibit DNA synthesis in herpesvirus-infected cells and for determining the inhibitory concentration of drug for virus isolates of the herpesvirus group.

4.3. Assay of Drug Sensitivity of Human Cytomegalovirus to Ganciclovir

The standard drug sensitivity test for HCMV is a time-consuming, labor-intensive plaque reduction assay.[24] The ability to use monoclonal antibodies to IE and late HCMV antigens and flow cytometry to determine the effect of antiviral drugs on the replication of HCMV suggested a rapid, quantitative procedure for determining the sensitivity of HCMV to ganciclovir. When cells are infected at an high Moi there is no drug-associated inhibition of the percentage of cells expressing the IE antigen even at drug concentrations that reduced the percentage of cells synthesizing the late antigen by greater than 90%. The ability to detect HCMV-infected cells expressing the IE antigen in the presence of inhibiting concentration of drugs is the basis for developing a flow cytometry-based drug sensitivity assay. MRC-5 cell monolayers were infected at an Moi of 10 with the drug-sensitive AD169 strain of HCMV or its ganciclovir-resistant derivative, D6/3/1, in the presence of various concentrations of ganciclovir. After incubation for 96 hr, the cells were harvested, treated with FITC-labeled monoclonal antibody to an IE antigen and PE-labeled monoclonal antibody to a late antigen, followed by RNAse and 7-AAD, and the cells were analyzed for three-color fluorescence by flow cytometry. The analysis was identical to that shown in Fig. 2. Figure 4 illustrates typical dot blots of FITC/GR-FL-positive cells analyzed for PE/OR-FL intensity versus FITC/GR-FL intensity showing the effect of inhibiting concentrations of ganciclovir on the synthesis of IE and late antigens in cells infected with AD169 or D6/3/1. Table II shows the data from these analyses. For the drug-sensitive AD169 strain, 1.5 μM and 12 μM ganciclovir have no effect on the percentage of cells synthesizing the IE antigen, but the percentage of cells synthesizing the late antigen is inhibited by approximately 50% and 95%, respectively. Even at concentrations of 12 μM ganciclovir, however, the percentage of cells synthesizing IE and late antigens is not blocked in cells infected with the drug-resistant D6/3/1 strain of HCMV. These results suggest that this assay can be used to distinguish between drug-sensitive and drug-resistant strains of HCMV.

Clinical isolates of HCMV are highly cell-associated. To determine the drug sensitivity of clinical isolates, cell-associated virus was added to monolayer cul-

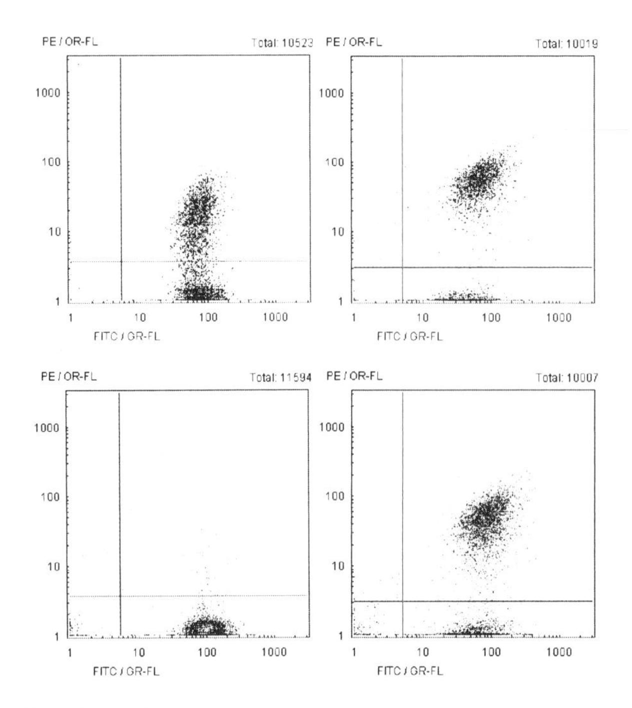

FIGURE 4. Effect of ganciclovir on IE and late-antigen synthesis in cells infected with drug-sensitive AD169 and drug-resistant D6/3/1 laboratory strains of HCMV. MRC-5 cells were infected with AD169 or D6/3/1 in the absence or presence of 12 μM ganciclovir. At 96 hr postinfection, the cells were harvested, prepared for flow cytometry, and analyzed as described in the legend in Fig. 2. The two upper panels show virally infected cells grown in the absence of ganciclovir. The two lower panels show virally infected cells grown in the presence of 12 μM ganciclovir.

tures of MRC-5 cells at various moi ranging from 0.1 to 1, and the cells were incubated in the presence of various concentrations of ganciclovir. After 96 hr of incubation, the cells were harvested and analyzed for three-color fluorescence by flow cytometry. The data in Table III show that for the ganciclovir-sensitive clinical isolate C9208, in the absence of ganciclovir, 25.2% of the cells express the IE antigen, and 10.4% of the cells express both the IE and late antigen, whereas at 12 μM ganciclovir only 3.3% of the cells express the IE antigen and 1.3% of the cells express both IE and late antigens. The IC_{50} of ganciclovir for this drug-sensitive isolate is 1.5 μM. For the ganciclovir-resistant clinical isolate C9209, the percentage of cell expressing the IE antigen or both the IE antigen and late antigen does not decrease with increasing concentrations of ganciclovir, yielding an IC_{50} of >12 μM. When five clinical isolates of unknown ganciclovir sensitivity were tested in this assay, clinical isolates 3 and 4 were sensitive to ganciclovir whereas clinical isolate 2 was resistant to ganciclovir and clinical isolate 5 was partially resistant to ganciclovir. Clinical isolate 1 had a substantial decline in the

TABLE II
Flow Cytometric Assay for Ganciclovir Sensitivity of HCMV Laboratory Strains

Drug (μM)	% IE AG+ cells	% Inhibition	% IE/late AG+ cells	% Inhibition	IC_{50}
0	98.8	0	40.8	0	
1.5	96.2	2.7	19.1	53	<1.5 μM
3.0	93.4	5.5	14.5	64	
12.0	96.4	2.5	2.2	95	

Ganciclovir-resistant D6/3/1-infected MRC-5 cells

Drug (μM)	% IE AG+ cells	% Inhibition	% IE/late AG+ cells	% Inhibition	IC_{50}
0	97.3	0	41.6	0	
1.5	97.0	0.3	38.8	6.7	
3.0	96.2	1.1	38.2	8.2	
12.0	94.7	2.6	41.7	0	>12 μM

TABLE III
Ganciclovir Sensitivities of HCMV Clinical Isolates

Virus	GCV μM	IE AG	IE/late AG	% Inhibition [IE AG]	% Inhibition [IE/late AG]	IC_{50}
C9208 (S)	0	25.2	10.4	0	0	1.5 μM
	1.5	12.4	5.0	51	52	
	12.0	3.3	1.3	87	88.5	
C9209 (R)	0	36.7	19.9	0	0	>12 μM
	1.5	36.2	21.0	1.4	5.2	
	12.0	34.1	19.0	7.1	4.6	
CLIN 1	0	27.0	18.0	0	0	>12 μM
	1.5	12.0	14.0	66	33	
	12.0	8.0	12.0	70	44	
CLIN 2	0	51.1	40.0	0	0	>12 μM
	1.5	58.0	40.5	0	0	
	12.0	54.1	36.9	0	0	
CLIN 3	0	25.0	23.0	0	0	>1.5 μM
	1.5	15.0	14.0	40	40	
	12.0	9.0	6.0	66	74	
CLIN 4	0	18.4	10.1	0	0	>1.5 μM
	1.5	11.7	7.9	47	21	
	12.0	2.0	2.5	90	86	
CLIN 5	0	13.8	9.4	0	0	>1.5 μM
	1.5	11.3	6.8	20	28	
	12.0	10.4	6.4	25	32	

percentage of cells synthesizing the IE antigen, but the percentage of cells synthesizing the IE and late antigen did not decline significantly. Thus, this clinical isolate had an IC_{50} of >12 μM. A major difference between the data derived from cell-free virus and that derived from cell-associated virus is the reduced percentage of cell expressing the IE antigen in the presence of inhibitory concentrations of ganciclovir. The experiments done with the cell-free virus used an Moi of 10 so that all of the cells would be infected at time 0 and second rounds of infection would be minimized. The experiments with the cell-associated virus were done with Moi of 0.1 to 1.0 where second rounds of infection were possible. At these lower Moi, the percentage of cells expressing both the IE and IE/late antigens was reduced because second rounds of infection were prevented in drug-sensitive clinical isolates in the presence of inhibitory concentrations of the drug. With a clear understanding of the effect of Moi on the assay, these results suggest that this drug sensitivity assay can be used for cell-associated HCMV clinical isolates.

5. FUTURE USES OF FLOW CYTOMETRY IN CLINICAL VIROLOGY

This review has demonstrated that fluorochrome-labeled antibodies and flow cytometry detect virally infected cells in culture and directly in clinical samples. There are at least four areas where flow cytometry can have an impact in the clinical virology laboratory. First, flow cytometry will be used for rapidly detecting virally infected cells directly in clinical samples. Most published examples of the direct detection of virally infected cells in blood of BAL used methanol to fix and permeabilize the cells. Methanol destroys the light-scattering properties of cells and some of the cell-surface antigens used to identify cells in clinical samples.[7] With new fixatives, such as Permeafix (Ortho Diagnostic Systems, Inc., Raritan, NJ), Fix and Perm (Caltag Laboratories, San Francisco, CA), and FACS Lysing Reagent and FACS Permeabilizing Solution (Becton Dickinson Co., San Jose, CA), which permeabilize cells without destroying light-scattering properties and surface antigens, it may be possible to detect virally infected cells directly in blood and BAL samples and to phenotype the infected cell.[25,26] As more fluorochrome-labeled monoclonal antibodies to viral antigens become available, the flow cytometric detection and quantitation of virally infected cells in clinical samples will become the method of choice for rapid laboratory diagnosis. This change will help eliminate labor-intensive and subjective methods involving fluorescence microscopy. With automation, detecting virally infected cells directly in clinical samples will add speed and efficiency to the clinical virology laboratory so that diagnostic virology will become a practical adjunct to clinical diagnosis. Second, the use of flow cytometry to detect and quantitate virally infected cells in culture will add speed and automation to the clinical laboratory. In the future, technicians will not spend many hours at the microscope looking for CPE or fluorescence. As a screen for antiviral drugs, flow cytometry will replace the standard plaque assay. Third, the role of apoptosis in viral infection is just beginning to be examined.[27] Flow cytometry will add quantitation to the currently used microscopic observations and by labeling the cell surface with antibodies to cell-surface antigens, the phenotype of the apoptotic cell can be determined. Fourth, flow cytometry will eventually be used to quantitate virally infected cells that have had either viral RNA or DNA amplified by PCR procedures. In the near future this promising technology will become practical in the clinical laboratory resulting in greater efficiency and lower costs.

Acknowledgments. The author acknowledges Ms. Christine Chutkowski, Ms. Ann Ogden-McDonough, and Ms. Betty Olson for their excellent work while performing the experiments presented in this chapter. This work was supported by grants AI30883 and AI32367 from the National Institutes of Health.

REFERENCES

1. Elmendorf, S., McSharry, J., Laffin, J., Fogleman, D., and Lehman, J., 1988, Detection of an early cytomegalovirus antigen with two color quantitative flow cytometry, *Cytometry* **9:**254–260.
2. McSharry, J., Costantino, R., Robbiano, E., Echols, R., Stevens, R., and Lehman, J., 1990, Detection and quantitation of human immunodeficiency virus-infected peripheral blood mononuclear cells by flow cytometry, *J. Clin. Microbiol.* **28:**724–733.
3. Ohlssom-Wilhelm, B., Cory, J., Kessler, H., Eyster, M., Rapp, F., and Landay, A., 1990, Circulating human immunodeficiency (HIV) p24 antigen-positive lymphocytes: A flow cytometric measure of HIV infection, *J. Infect. Dis.* **162:**1018–1024.
4. Costigliola, P., Tumietto, F., Ricchi, E., and Chiodo, F., 1992, Detection of circulating p24 antigen-positive cells during HIV infection by flow cytometry, *AIDS* **6:**1121–1125.
5. Holzer, T., Heynen, C., Novak, R., Pitrak, D., and Dawson, G., 1993, Frequency of cells positive for HIV-p24 antigen assessed by flow cytometry, *AIDS* **7:**S2–S5.
6. Gadol, N., Crutcher, G., and Bush, M., 1994, Detection of intracellular HIV in lymphocytes by flow cytometry, *Cytometry* **15:**359–370.
7. McSharry, J., 1994, Uses of flow cytometry in virology, *Clin. Microbiol. Rev.* **7:**576–604.
8. Shapiro, H. M., 1995, *Practical Flow Cytometry*, 3rd ed., Wiley-Liss, New York.
9. van der Bij, W., Schirm, J., Torensma, R., van Son, W., Tegzess, A., and The, T., 1988, Comparison between viremia and antigenemia for detection of cytomegalovirus in blood, *J. Clin. Microbiol.* **26:**2531–2535.
10. Grefte, J., van der Gun, B., Schmolke, S., van der Geissen, M., van Son, W., Plachter, B., and The, T., 1992, The lower matrix protein pp65 is the principal viral antigen present in peripheral blood leukocytes during active cytomegalovirus infection, *J. Gen. Virol.* **73:**2923–2932.
11. Gerna, G., Revello, M., Percivalle, E., and Morini, F., 1992, Comparison of different immunostaining techniques and monoclonal antibodies to the lower matrix phosphoprotein (pp65) for optimal quantitation of human cytomegalovirus antigenemia, *J. Clin. Microbiol.* **30:**1232–1237.
12. Erice, A., Holm, M., Gill, P., Henry, S., Dirksen, C., Dunn, D., Hillman, R., and Balfour, H., Jr., 1992, Cytomegalovirus (CMV) antigenemia assay is more sensitive than shell vial cultures for

rapid detection of CMV in polymorphonuclear blood leukocytes, *J. Clin. Microbiol.* **30:**2822–2825.

13. Roizman, B. and Sears, A., 1995, Herpes simplex viruses and their replication, in: *Fields Virology*, (B. N. Fields, D. M. Knipe, and P. M. Howley, eds.). Lippincott-Raven Press, New York, Vol. 2, pp. 2231–2296.
14. McSharry, J., Costantino, R., McSharry, M., Venezia, R., and Lehman, M., 1990, Rapid detection of herpes simplex virus in clinical samples by flow cytometry after amplification in tissue culture, J. Clin. Microbiol. **28:**1864–1866.
15. Schols, D., Snoeck, R., Neyts, J., and De Clercq, E., 1989, Detection of immediate early, early, and late antigens of human cytomegalovirus by flow cytometry, *J. Virol. Methods.* **26:**247–254.
16. Chutkowski, C. and McSharry, J., Flow cytometric analysis of antigen expression in human cytomegalovirus (HCMV) infected cells. *13th Annual Meeting of the American Society for Virology,* Madison, WI, July 9–13, 1994.
17. McSharry, J. and Chutkowski, C., Rapid assay for determining the mode of action of compounds effective against human cytomegalovirus (HCMV), *International Society for Antiviral Research,* Sante Fe, NM, April 23–28, 1995.
18. McSharry, J., Lurain, N., Manichewitz, J., Notka, M., and Weinberg, A., Standardized flow cytometry based assay for HCMV antiviral susceptibility. *3rd Conference on Retroviruses and Opportunistic Infections,* Washington, DC, Jan. 28–Feb. 2, 1996.
19. McSharry, J., Lurain, N., Notka, M., O'Gorman, M., Shapiro, H., and Weinberg, A., A rapid, quantitative assay for ganciclovir resistant human cytomegalovirus (HCMV) clinical isolates, *4th Conference on Retroviruses* and Opportunistic Infections, Washington, DC, January 22–27, 1997.
20. Lurain, N., Thompson, K., Holmes, E., and Read, G., 1992, Point mutations in the DNA polymerase gene of human cytomegalovirus that result in resistance to antiviral agents, *J. Virol.* **66:**7146–7152.
21. Hirsch, M., Kaplan, J., and D'Aquila, R., 1995, Antiviral agents, in *Fields Virology* (B. M. Fields, D. M. Knipe, and P. M. Howley, eds.). Lippincott-Raven Press, New York, Vol. 1, pp. 431–466.
22. Mocarski, E., 1995, Cytomegaloviruses and their replication in: *Fields Virology,* (B. N. Fields, D. M. Knipe, and P. M. Howley, eds.), Lippincott-Raven Press, New York, pp. 2447–2492.
23. McSharry, J. J., 1995, Flow cytometry-based antiviral resistance assays. *Clin. Immunol. Newsl.* **15:**113–119.
24. Stanat, S., Reardon, J., Erice, A., Drew, W., and Biron, K., 1991. Ganciclovir-resistant clinical isolates: Mode of resistance to ganciclovir, *Antimicrob. Agents Chemother.* **35:**2191–2197.
25. Francis, C. and Connelly, M. C., 1996, Rapid single-step method for flow cytometric detection of surface and intracellular antigens using whole blood. *Cytometry* **25:**58–70.
26. Syrjala, M. T., Tiirikainen, T., Jansson, S-E., and Krusius, T., 1993, Flow cytometric analysis of terminal deoxynucleotide transferase: A simplified method, *Am. J. Clin. Path.* **99:**298–303.
27. Ramiro-Ibanez, F., Ortega, A., Brun, A., Escribano, J., and Alonso, C., 1996, Apoptosis: A mechanism of cell killing and lymphoid organ impairment during acute African swine fever virus infection, *J. Gen. Virol.* **77:**2209–2219.

4

Diagnostic Significance of Antibodies in Oral Secretions

SPENCER R. HEDGES, MICHAEL W. RUSSELL, and JIRI MESTECKY

1. INTRODUCTION

Immunoglobulins (Ig) were identified in human saliva nearly 40 years ago,[1] and shortly thereafter in 1963 the prevalence of IgA in saliva was demonstrated.[2] These pioneering studies prompted subsequent investigations into various aspects of salivary Ig, such as the presence of Ig isotypes in various salivary secretions, the structure, function, and origin of salivary Ig, and effective immunization routes for inducing salivary immune responses (reviews[3,4]).

Several notable differences between the mucosal and systemic compartments of the immune system reflect the different functional requirements of immune responses at mucosal sites where tissue integrity must be maintained in the presence of potentially pathogenic organisms. The predominant Ig type in human mucosal secretions, including saliva, is IgA in the form of secretory IgA (S-IgA) rather than IgG, which predominates in serum. In addition, salivary glands (and other glands and mucosal organs) are infiltrated by IgA-producing plasma cells, and the ductal and acinar epithelial cells express receptors specific for polymeric IgA molecules involved in the selective transport of IgA into saliva. The known links between different sites of mucosal immunity (the common mucosal immune system) [CMIS] suggest that the analysis of immune responses in saliva

SPENCER R. HEDGES, MICHAEL W. RUSSELL, and JIRI MESTECKY • Department of Microbiology, University of Alabama at Birmingham, Birmingham, Alabama 35294-2170.

Rapid Detection of Infectious Agents, edited by Specter *et al.* Plenum Press, New York, 1998.

may provide a general picture of the entire mucosal immune system. However, critical interpretation of salivary antibodies for diagnostic purposes requires some understanding of the mucosal immune system and how it differs from the circulatory immune system and the origins of salivary antibodies.

2. IgA AND THE MUCOSAL IMMUNE SYSTEM

IgA is the most heterogeneous human immunoglobulin.[5] It occurs in two subclasses, IgA1 and IgA2, which differ by 20 amino acid residues in the constant regions of $\alpha 1$ and $\alpha 2$ heavy chains, but there are also more N-linked glycan chains in IgA2 (review[6]). In addition, the heavy-chain hinge regions are very different. That of IgA1 contains an eight-residue proline-rich repeat susceptible to cleavage by unique IgA1-specific proteases produced by several bacterial species (review[7]), and furthermore contains 5 O-linked glycans, whereas the IgA2 hinge is much shorter, lacks glycosylation, and is not cleavable by IgA1 proteases.

IgA exists in all mammalian species in three distinct molecular forms that are differentially distributed between the systemic and mucosal compartments. Monomeric IgA (mIgA) contains two heavy (α) and two light (κ or λ) chains linked by covalent (disulfide) and noncovalent bonds. Polymeric IgA (pIgA) is composed of two or more disulfide-linked monomers and an additional small glycoprotein, the J chain. Serum IgA is predominantly monomeric and has only a small proportion (1–10%) of pIgA. In contrast, IgA in external secretions of the gastrointestinal, respiratory, and genitourinary tracts and salivary, lacrimal, and mammary glands is mostly polymeric and consists of dimers with smaller amounts of tetramers. A large, carbohydrate-rich glycoprotein, secretory component (SC; M_R ~80,000) becomes disulfide-linked to pIgA during the selective transport through epithelial cells into external secretions (reviews[3,4,8]). Small amounts of mIgA that are usually found in external secretions are derived from the plasma by passive diffusion, but pIgA present in plasma is not effectively transported into saliva.[9] Nearly all of the IgA in secretions is produced locally by pIgA-secreting cells resident in the submucosa or glandular stroma and is actively transported by SC into the secretion.

The distribution of Ig in human external secretions, such as saliva, is very different from that found in serum. IgA, mainly in the form of secretory IgA (S-IgA), is the predominant Ig type in mucosal secretions, whereas IgG dominates in serum. The presence of monomeric IgA, IgG, IgD, and IgE in external secretions is primarily due to passive transudation rather than active transport of plasma-derived Ig.[4] Several cytokines regulate SC expression on the surfaces of epithelial cells.[8,10,11] Interferon-γ (IFN-γ), interleukin (IL)-4, and tumor necrosis factor-α (TNF-α) enhance the expression of membrane SC either alone or synergistically.[12] These cytokines, which are produced by T cells and other cells in the mucosal tissues, therefore may regulate expression of SC (and thereby S-IgA secretion).

Locally present mucosal cytokines also profoundly influence the maturation, differentiation, isotype commitment, and Ig secretion of lymphocytes of the B cell lineage, including those involved in IgA synthesis.[13] Mucosal epithelial cells produce a variety of cytokines in response to physiological stimuli, including IL-6 and transforming growth factor-β (TGF-β), both of which are involved in differentiating B cells into IgA secretors (review[14]). Furthermore, the local synthesis of IL-6 by epithelial rather than T cells is significantly up-regulated as a consequence of local infection.[15,16] Moreover, mucosal epithelial cells also produce a variety of cytokine receptors suggesting that they participate in mucosal cytokine networks.[14,17–19] Thus, infection-induced IL-6 and possibly TGF-β synthesis, in turn, may promote the differentiation of IgA-producing plasma cells to enhance the secretion of S-IgA.

The mucosal plasma cells that secrete pIgA are now known to originate in the inductive sites of mucosal immunity, of which the best known are the Peyer's patches (PP) of the small intestine, but include an addition analogous lymphoid follicles in the large bowel, respiratory tract, and Waldeyer's ring.[3,4] Therefore the concept of a CMIS has arisen: antigens are taken up by specialized epithelial cells (M cells) covering the lymphoid follicles and enter the underlying populations of antigen-processing and antigen-presenting cells. Immature surface IgM$^+$ (sIgM$^+$) B cells switch to become sIgA$^+$ under the influence of cytokines from local T cells (reviews[13]). Such isotype- and antigen-committed cells emigrate and after maturation in mesenteric (or corresponding) lymph nodes, enter the blood circulation through the thoracic duct, and lodge in the mucosal tissues and glands where terminal differentiation into pIgA-producing plasma cells occurs. Although the evidence is necessarily largely indirect, many experiments strongly indicate that the CMIS operates in humans (review[3,20]). Despite the commonality of mucosal immunity that arises from the dissemination of T and B cells from inductive to effector sites, increasing evidence suggests that the distribution is not uniform, but that instead some compartmentalization occurs. Thus, responses induced through PP may be preferentially (but not exclusively) expressed in the intestine, whereas those induced through the Waldeyer's ring tissues (or equivalent nasal lymphoid tissue in rodents) are preferentially expressed in the upper respiratory tract, saliva, and (interestingly) in the genital tract, and rectal immunization may also selectively enhance responses in the lower bowel and genital tract. However, the mechanisms responsible for the distribution of these responses are not understood. Nevertheless, an important consequence is that determining antibody responses in one secretion, such as saliva, may not be truly representative of all mucosal responses.

TABLE I
Immunoglobulin Concentrations (Mean ±SD, μg/ml) in Human Salivary Secretions[a]

Secretion	IgA	IgG	IgM
Parotid saliva (stimulated)	39.5 ± 13.7	0.36 ± 0.30	0.43 ± 0.36
Parotid saliva (resting)	119.6 ± 48.3	nd	nd
Whole saliva (resting) (normal adults)	194.0 ± 53.7	14.4 ± 9.0	2.1 ± 1.9
Whole saliva (resting) (periodontitis patients)	371.4 ± 224.7	69.7 ± 33.6	7.6 ± 5.4

[a]Data from Ref.[4].

3. ORIGINS OF SALIVARY IMMUNOGLOBULINS

The great convenience of using saliva as a means of assessing mucosal immune responses lies in its obvious accessibility. However, it must be borne in mind that human saliva is a mixture of secretions from three major sets of glands (parotid, submandibular, and sublingual), plus the minor salivary glands distributed particularly on the inside of the lower lip,[21] and a small but highly variable contribution of gingival crevicular fluid.[4] The latter is a transudate of tissue fluid and plasma, and its volume, which varies greatly according to the state of gingival health, becomes greater as gingivitis and periodontal disease increase in severity.[22] Furthermore, in advanced periodontal disease, there is an accumulation of B cells and their progeny in the gingiva, which secrete Ig that probably reflects locally induced antibodies against the periodontal microbiota.[23] Even in the normally healthy adult human, gingival crevicular fluid is responsible for most of the IgG in whole saliva,[24] because the pure glandular secretions contain substantially lower concentrations.[4] Lesions in the oral mucosa also lead to increased transudation of tissue fluid. A further potential complication arising from analyzing whole saliva is that exposure to organisms within the oral cavity results in adsorption of antibodies.[25] or their degradation by proteases, including IgA1-specific proteases produced by a number of oral bacteria.[7] Typical Ig concentrations in human salivary secretions are shown in Table I.

4. DIAGNOSIS OF INFECTION AND DISEASE WITH SALIVARY ANTIBODIES

Given the links between the different elements of the CMIS, it may be possible to diagnose mucosal infections based on detecting specific antibodies in mucosal secretions.[3] Saliva is the most easily accessible mucosal secretion. All others require an invasive procedure for collection. Saliva is often used for analyzing mucosal immune responses to experimental vaccines. For example, intranasal infection with attenuated influenza virus induces pronounced immune responses manifested by the presence of IgA and IgG antibodies in nasal secretions, saliva, and serum of volunteers.[26] Furthermore, the ingestion of formalin-inactivated influenza virus produces predominantly IgA virus-specific antibodies in saliva,[27] whereas systemic immunization induces excellent serum IgG antibodies, but only weak or no salivary IgA responses. Similarly, although systemic immunization with bacterial capsular polysaccharides (from *Streptococcus pneumoniae, Haemophilus influenzae,* or *Neisseria meningitidis,* as free polysaccharides or as protein-conjugated complexes) induces excellent serum antibodies, including pIgA2, these are not transported to saliva or reflected in the salivary gland responses.[28] However, ingestion of or rectal immunization with live *Salmonella typhi* Ty21a vaccine induces specific IgA antibodies in the saliva,[29,30] and repeated intrarectal immunization with formalin-inactivated influenza virus induces both IgA and IgG antibodies in saliva.[31]

It might be thought that salivary IgA antibodies would be particularly prevalent in oral infections, such as dental caries, periodontal disease, and oral candidiasis (thrush), and hence reflect disease activity. An extensive literature exists on detecting salivary antibodies to the causative organisms of caries, the 'mutans streptococci', and their antigens but with variable and conflicting conclusions. As a result, no criteria useful in diagnosing or predicting immunity can be delineated currently.[32] In periodontal disease, salivary IgA antibodies to various putative periodontal pathogens may be quite abundant, but their diagnostic or prognostic potential has not been established.[32,33] Recent findings suggest that elevated concentrations of IgA in the gingival pocket fluid of periodontal disease patients are associated with a more favorable response to conventional treatment.[34] However, this IgA originates from plasma and tissue fluid and not the salivary secretions, and furthermore, information about specific antibodies to periodontal pathogens is not available currently. Salivary IgA antibodies to *Candida albicans* are also increased in this infection and may have protective value,[35] but their elevation is not used diagnostically. Although, criteria of infection more overt than assay of salivary antibodies are often possible in oral diseases, conceivably salivary anti-*Candida* antibodies could help in detecting genital tract or systemic infection with this organism. Although HIV-infected individuals, who are susceptible to *C. albicans* infection, display diminished salivary antibodies to candidal antigen, this is considered insufficient to account for such susceptibility.[35]

The HIV epidemic has given great impetus to the diagnostic potential of determining salivary antibodies. Although serological identification of antibodies to HIV in blood serum is the most widely used diagnostic method, the detection of salivary antibodies in HIV-infected subjects[36–38] has led to numerous reports examining the potential for diagnosing HIV infection based on anti-HIV antibodies in saliva.[39–43] It has been concluded that analysis of saliva accurately diagnoses HIV infection, and the first commercial test was approved for trial by the United States Food and Drug Administration in 1990. The advantages of saliva-based rather than serum-based tests include faster and less expensive diagnosis and increased safety. Interestingly, in the saliva of HIV-infected individuals, including those who acquire it heterosexually, almost all HIV-specific antibodies are IgG, and only rarely of the IgA isotype.[44] Conversely, salivary IgA antibodies to HIV have been found in infected infants and children, and their determination has been proposed as a diagnostic aid in infants under 15 months of age.[41] However, specific IgA antibodies have not been convincingly detected in parotid saliva samples collected from volunteers *systemically* immunized with experimental HIV vaccines (Mestecky *et al.*, unpublished results).

Determination of salivary antibodies to hepatitis viruses A, B, and C is a sensitive and useful procedure, particularly in children or other situations where blood sampling is difficult.[45–48] Although these procedures detect either IgM or IgG antibodies, presumably originating from the plasma, a method for detecting salivary S-IgA antibodies to hepatitis B antigen has been reported.[49] Salivary antiviral antibodies, particularly IgM, have been detected in the majority of cases of measles, mumps, and rubella within one to five weeks of the onset of symptoms.[50] IgG antibodies were less often detected, and IgA antibodies were not reported. Among other viral infections, salivary IgG antibodies to cytomegalovirus are associated with infection in preschool children,[51] and the diagnosis of parvovirus B19 infection has been facilitated by salivary IgM or IgG antibody detection.[52,53]

The diagnosis of mucosal bacterial infections by assaying salivary antibodies has been less successful. Several salivary IgG antibody tests for *Helicobacter pylori* infection have been developed,[54–58] but generally they are less accurate than serum-based assays. Antibodies to the lipopolysaccharide of *Shigella sonnei* and *Shigella flexneri* have been detected in the saliva of infected patients,[59] but a lack of species specificity precluded their efficient use in diagnosis. Likewise, it has been concluded that, although patients infected with *Bordetella pertussis* often develop salivary IgA antibodies to pertussis toxin and the filamentous hemagglutinin within 6 to 50 days after the onset of symptoms, the titer increase was not sufficiently reliable or prompt enough for diagnostic purposes.[60] Salivary antibodies to *Leptospira,* however, show promise as a useful diagnostic tool for leptospirosis.[61]

In the parasite field, salivary IgG antibodies to *Trichuris trichiura* are significantly elevated in infected children, whereas those to *Ascaris lumbricoides* and hookworms, though elevated, do not attain statistical significance.[62] Nevertheless, such assays are thought to have epidemiological value. A few reports also describe salivary antibody assays for detecting antibodies to *Leishmania,*[63] *Toxoplasma gon-*

dii,[64] and *Entamoeba histolytica*.[65] In the last instance, salivary IgA antibodies have a diagnostic accuracy of >90%, and a correlation with fecal antibody was noted.

In addition to diagnosing mucosal and other infections, salivary antibodies may also be helpful in diagnosing diseases of immune hypersensitivity or autoimmune etiology. For example, salivary antigliadin and antiendomysium antibodies are as useful (if not better) as serum antibodies in a screening test for celiac disease.[66] The diagnosis of Sjögren's syndrome, an autoimmune condition that involves the salivary glands, may be facilitated by determining salivary IgG concentrations, which are elevated according to lymphocytic infiltration of the salivary glands.[67,68]

5. PRACTICALITIES AND LIMITATIONS

It is important to recognize that the concentration of IgA (and other Igs) in saliva varies inversely with the flow rate of the secretions.[69–71] Salivary flow is subject to wide short-term fluctuations according to extrinsic (e.g., gustatory) stimuli and longer term variation due to intrinsic stimuli (including circadian rhythms) acting via the neuroendocrine systems. The data in Table I illustrate the considerable variability in salivary Ig concentrations and a great difference between stimulated and resting saliva. Thus it appears that synthesis of pIgA by plasma cells within the glandular stroma and its transport through the epithelia are relatively constant, but increased fluid secretion results in corresponding dilution of the S-IgA. Furthermore, individuals vary markedly in salivary IgA concentration, and the same individual can show long-term changes.[71,72] For all of these reasons, therefore, it is important that specific antibody levels be related to the concentration of total Ig of the corresponding isotype. Fluctuations in the latter will obviously affect the former.

For practical reasons, it is often easier to obtain stimulated rather than "resting" saliva. This is most readily accomplished by applying citric acid solution (or citrus fruit juice) to the tongue tip. Mechanical stimulation, for example, by chewing paraffin wax or rubber, is less satisfactory because this increases the dislodgement of bacteria, food debris, and epithelial cells, and also promotes gingival crevicular fluid exudation.[73] Collection of parotid saliva is preferable because this represents a "pure" secretion uncontaminated with other oral fluids and not exposed to the potentially degradative and absorptive effects of the oral cavity and its contents. Furthermore, parotid saliva is serous rather than mucous in character and therefore easier to handle. The best method of collecting parotid saliva is by using Curby cups[74] (or similar devices), which are placed over the openings of Stensen's ducts and held there by vacuum applied through the outer ring of the cup, while the fluid drains through the central channel into collecting tubes. A simpler alternative is the Schaefer cup,[75] which is placed into the buccal sulcus to collect the flow from Stensen's duct. In both cases, citric acid stimulus can be applied to the tongue without contaminating the secretion. Collecting the output of the submandibular and sublingual glands is more difficult and requires isolating (or diverting) parotid secretions and removing the fluid that accumulates under the tongue by a pipet. Because these secretions are very mucinous, however, they are less suitable than parotid saliva for immunochemical assays. The secretions of the minor labial glands are not amenable to routine collection because only minute volumes of very viscous fluid are produced.

6. CONCLUSIONS

Salivary antibodies can be used to diagnose a variety of diseases when the relevant specific antigens are known. Given the functional separation between the mucosal and systemic immune systems, this may be especially useful for diagnosing mucosal infections. Presently, however, this capability is still mostly in the developmental stage, and it is remarkable that the majority of assays described so far detect salivary IgM or IgG antibodies which probably originate in the circulatory rather than the mucosal immune system. From reports published to date, salivary antibody is most successful when used in diagnosing viral infections, but why the diagnosis of bacterial infections should be more problematic is not clear. Many of the mechanisms underlying the regulation and distribution of mucosal immunity, however, remain to be elucidated. Further understanding of the pathogenesis of bacterial mucosal infections may enable their diagnosis by salivary antibodies. An obvious market exists for products based on this methodology for diagnosing HIV and other newly emerging infections, such as *H. pylori*. Realizing the potential benefits of diagnosis based on salivary antibodies, necessitates further research and clearer understanding of the relationship of salivary secretions to the mucosal immune system and the origins of salivary antibodies.

Acknowledgments. Studies in the authors' laboratories are supported by USPHS grants DE06746, DE 08228, DE 08182, DE09691, AI 28147, AI 35163, and AI 34970.

REFERENCES

1. Ellison, S. A., Mashimo, P. I., and Mandel, I. D., 1960, Immunochemical studies of human saliva. I. The demonstration of serum proteins in whole and parotid saliva, *J. Dent. Res.* **39:**892–898.
2. Tomasi, T. B. and Ziegelbaum, S., 1963, The selective occurrence of γ1A globulins in certain body fluids, *J. Clin. Invest.* **42:**1552–1560.

3. Mestecky, J. and McGhee, J. R., 1987, Immunoglobulin A (IgA): Molecular and cellular interactions involved in IgA biosynthesis and immune response, *Adv. Immunol.* **40:**153–245.
4. Brandtzaeg, P., 1989, Salivary immunoglobulins, in: *Human Saliva: Clinical Chemistry and Microbiology*, (J. O. Tenovuo, ed.). CRC Press, Boca Raton, Florida, Vol. II, pp. 1–54.
5. Russell, M. W., Lue, C., van den Wall Bake, A. W. L., Moldoveanu, Z., and Mestecky, J., 1992, Molecular heterogeneity of human IgA antibodies during an immune response, *Clin. Exp. Immunol.* **87:**1–6.
6. Mestecky, J. and Russell, M. W., 1986, IgA Subclasses, Monogr. Allergy **19:**277–301.
7. Kilian, M., Reinholdt, J., Lomholt, H., Poulsen, K., and Frandsen, E. V. G., 1996, Biological significance of IgA1 proteases in bacterial colonization and pathogenesis: Critical evaluation of experimental evidence, *APMIS* **104:**321–338.
8. Mestecky, J., Lue, C., and Russell, M. W., 1991, Selective transport of IgA: cellular and molecular aspects, *Gastroenterol. Clin. N. Amer.* **20:**441–471.
9. Kubagawa, H., Bertoli, L. F., Barton, J. C., Koopman, W. J., Mestecky, J., and Cooper, M. D., 1987, Analysis of paraprotein transport into saliva by using anti-idiotype antibodies, *J. Immunol.* **138:**435–439.
10. Phillips, J. O., Everson, M. P., Moldoveanu, Z., Lue, C., and Mestecky, J., 1990, Synergistic effect of IL-4 and IFN-γ on the expression of polymeric Ig receptor (secretory component) and IgA binding by human epithelial cells, *J. Immunol.* **145:**1740–1744.
11. Kvale, D., Brandtzaeg, P., and Løvhaug, D., 1988, Up-regulation of the expression of secretory component and HLA molecules in a human colonic cell line by tumor necrosis factor-α and gamma interferon, *Scand. J. Immunol.* **28:**351–357.
12. Denning, G. M., 1996, IL-4 and IFN-γ synergistically increase total polymeric IgA receptor levels in human intestinal epithelial cells. Role of protein tyrosine kinases, *J. Immunol.* **156:** 4807–4814.
13. McGhee, J. R., Mestecky, J., Elson, C. O., and Kiyono, H., 1989, Regulation of IgA synthesis and immune response by T cells and interleukins, *J. Clin. Immunol.* **9:**175–199.
14. Hedges, S. R., Agace, W. W., and Svanborg, C., 1995, Epithelial cytokine responses and mucosal cytokine networks, *Trends Microbiol.* **3:**266–270.
15. Hedges, S., Linder, H., de Man, P., and Svanborg Eden, C., 1990, Ciclosporin-dependent, *nu*-independent, mucosal interleukin-6 response to gram-negative bacteria. *Scand. J. Immunol.* **31:** 335–343.
16. Hedges, S., Anderson, P., Lidin-Janson, G., de Man, P., and Svanborg, C., 1991, Interleukin-6 response to deliberate colonization of the human urinary tract with gram-negative bacteria, *Infect. Immun.* **59:**421–427.
17. Shirota, K., LeDuy, L., Yuan, S., and Jothy, S., 1990, Interleukin-6 and its receptor are expressed in human intestinal epithelial cells, *Virchow's Archiv. B, Cell. Pathol. Molec. Pathol.* **58:**303–308.
18. Ciacci, C., Mahida, Y. R., Dignass, A., Koizumi, M., and Podolsky, D. K., 1993, Functional interleukin-2 receptors on intestinal epithelial cells, *J. Clin. Invest.* **92:**527–532.
19. Arnold, R., Humbert, B., Wechau, H., Gallati, H., and König, W., 1994, Interleukin-8, interleukin-6, and soluble tumour necrosis factor receptor type I release from a human pulmonary epithelial cell line (A549) exposed to respiratory syncytial virus, *Immunology* **82:**126–133.
20. Mestecky, J., 1987, The common mucosal immune system and current strategies for induction of immune response in external secretions, *J. Clin. Immunol.* **7:**265–276.
21. Crawford, J. M., Taubman, M. A., and Smith, D. J., 1975, Minor salivary glands as a major source of immunoglobulin A in the human oral cavity, *Science,* **190:**1206–1209.
22. Ranney, R. R., Ruddy, S., Tew, J. G., Welsheimer, H. J., Palcanis, K. G., and Segreti, A., 1981, Immunological studies of young adults with severe periodontitis. I. Medical evaluation and humoral factors, *J. Periodont. Res.* **16:**390–402.
23. Seymour, G. J., Powell, R. N., and Davis, W. I. R., 1979, Conversion of a stable T-cell lesion to a

progressive B-cell lesion in the pathogenesis of chronic inflammatory disease. A hypothesis, *J. Clin. Periodontol.* **6:**2672–2677.
24. Challacombe, S. J., Russell, M. W., and Hawkes, J. E., 1978, Passage of intact IgG from plasma to the oral cavity via crevicular fluid, *Clin. Exp. Immunol.* **34:**417–422.
25. Brandtzaeg, P., Fjellanger, I., and Gjeruldsen, S. T., 1968, Adsorption of immunoglobulin A onto oral bacteria *in vivo, J. Bacteriol.* **96:**242–249.
26. Moldoveanu, Z., Clements, M. L., Prince, S. J., Murphy, B. R., and Mestecky, J., 1995, Human immune responses to influenza virus vaccines administered by systemic or mucosal routes, *Vaccine* **13:**1006–1012.
27. Bergmann, K. C. and Waldmann, R. H., 1988, Stimulation of secretory antibody following oral administration of antigen, *Rev. Infect. Dis.* **10:**939–950.
28. Tarkowski, A., Lue, C., Moldoveanu, Z., Kiyono, H., McGhee, J. R., and Mestecky, J., 1990, Immunization of humans with polysaccharide vaccines induces systemic, predominantly polymeric-IgA2 subclass antibody responses, *J. Immunol.* **144:**3770–3778.
29. Forrest, B. D., Shearman, D. J. C., and LaBrooy, J. T., 1990, Specific immune responses in humans following rectal delivery of five typhoid vaccine, *Vaccine* **8:**209–212.
30. Kantele, A. and Mäkelä, P. H., 1991, Different profiles of the human immune response to primary and secondary immunization with an oral *Salmonella typhi* Ty21a vaccine, *Vaccine* **9:**423–427.
31. Crowley-Nowick, P. A., Bell, M. C., and Brockwell, R., Edwards, R. P., Chen, S., Partridge, E. E., and Mestecky, J., 1997, Rectal immunization for induction of specific antibody in the genital tract of women *J. Clin. Immunol.* **17:**370–379.
32. Taubman, M. A. and Smith, D. J., 1993, Significance of salivary antibody in dental diseases, *Ann. N.Y. Acad. Sci.* **694:**202–215.
33. Wilton, J. M. A., Johnson, N. W., Curtis, M. A., Gillett, I. R., Carman, R. J., Bampton, J. L. M., Griffiths, G. S., and Sterne, J. A. C., 1991, Specific antibody responses to subgingival plaque bacteria as aids to the diagnosis and prognosis of destructive periodontitis, *J. Clin. Periodontol.* **18:**1–15.
34. Grbic, J. T., Singer, R. E., Jans, H. H., Celenti, R. S., and Lamster, I. B., 1995, Immunoglobulin isotypes in gingival crevicular fluid: Possible protective role of IgA, *J. Periodontol* **66:**55–61.
35. Coogan, M. M., Sweet, S. P., and Challacombe, S. J., 1994, Immunoglobulin A (IgA), IgA1, and IgA2 antibodies to *Candida albicans* in whole and parotid saliva in human immunodeficiency virus infection and AIDS, *Infect. Immun.* **62:**892–896.
36. Archibald, D. W., Barr, C. E., Torosian, J. P., McLane, M. F., and Essex, M., 1987, Secretory IgA antibodies to human immunodeficiency virus in the parotid saliva of patients with AIDS and AIDS-related complex, *J. Infect. Dis.* **155:**793–796.
37. Parry, J. V., Perry, K. R., and Mortimer, P. P., 1987, Sensitive assays for viral antibodies in saliva: An alternative to tests on serum, *Lancet* **ii:**72–75.
38. Funkhouser, A., Clements, M. L., Slome, S., Clayman, B., and Viscidi, R., 1993, Antibodies to recombinant gp160 in mucosal secretions and sera of persons infected with HIV-1 and seronegative vaccine recipients, *AIDS Res. Hum. Retroviruses* **9:**627–632.
39. Johnson, A. M., Parry, J. V., Best, S. J., Smith, A. M., de Silva, M., and Mortimer, P. P., 1988, HIV surveillance by testing saliva, *AIDS* **2:**369–371.
40. Holmström, P., Syrjanen, S., Laine, P., Valle, S. L., and Suni, J., 1990, HIV antibodies in whole saliva detected by ELISA and western blot assays, *J. Med. Virol.* **30:**245–248.
41. Archibald, D. W., Johnson, J. P., Nair, P., Alger, L. S., Hebert, C. A., Davis, E., and Hines, S. E., 1990, Detection of salivary immunoglobulin A antibodies to HIV-1 in infants and children, *AIDS* **4:**417–420.
42. Major, C. J., Read, S. E., Coates, R. A., Francis, A., McLaughlin, B. J., Millson, M., Shepherd, F., Fanning, M., Calzavara, L., McFadden, D., Johnson, J. K., and O'Shaughnessy, M. V., 1991,

Comparison of saliva and blood for human immunodeficiency virus prevalence testing, *J. Infect. Dis.* **163:**699–702.

43. Lu, X. S., Delfraissy, J. F., Grangeot-Keros, L., Rannou, M. T., and Pillot, J., 1994, Rapid and constant detection of HIV antibody response in saliva of HIV infected patients; selective distribution of anti-HIV activity in the IgG isotype, *Res. Virol.* **145:**369–377.
44. Casteel, L. R., Allen, S., Kulhavy, R., Zulu, I., Omara, J., and Mestecky, J., 1997, HIV-specific antibodies in external secretions of sero-negative and sero-positive high-risk individuals, 119th Ann. Meeting Nation. Coop. Vaccine Develop. Group AIDS. Abstr. No. 73, p. 178.
45. Sherman, K. E., Creager, R. L., O'Brien, J., Sargent, S., Piacentini, S., and Thieme, T., 1994, The use of oral fluid for hepatitis C antibody screening, *Am. J. Gastroenterol.* **89:**2025–2027.
46. Thieme, T., Yoshihara, P., Piacentini, S., and Beller, M., 1992, Clinical evaluation of oral fluid samples for diagnosis of viral hepatitis, *J. Clin. Microbiol* **30:**1076–1079.
47. Stuart, J. M., Majeed, F. A., Cartwright, K. A., Room, R., Parry, J. V., Perry, K. R., and Begg, N. T., 1992, Salivary antibody testing in a school outbreak of hepatitis A, *Epidemiol. Infect.* **109:**161–166.
48. Parry, J. V., Perry, K. R., Panday, S., and Mortimer, P. P., 1989, Diagnosis of hepatitis A and B by testing saliva, *J. Med. Virol.* **28:**255–260.
49. Siddiqi, M. A., and Abdullah, S., 1988, An 'antigen capture' ELISA for secretory immunoglobulin A antibodies to hepatitis surface antigen in human saliva, *J. Immunol. Methods.* **114:**207–211.
50. Perry, K. R., Brown, D. W., Parry, J. V., Panday, S., Pipkin, C., and Richards, A., 1993, Detection of measles, mumps, and rubella antibodies in saliva using antibody capture radioimmunoassay, *J. Med. Virol.* **40:**235–240.
51. Wang, J. B., and Adler, S. P., 1996, Salivary antibodies to cytomegalovirus (CMV) glycoprotein B accurately predict CMV infections among preschool children, *J. Clin. Microbiol.* **34:**2632–2634.
52. Cubel, R. C., Oliviera, S. A., Brown, D. W., Cohen, B. J., and Nascimento, J. P., 1996, Diagnosis of parvovirus B19 infection by detection of specific immunoglobulin M antibody in saliva, *J. Clin. Microbiol.* **34:**205–207.
53. Rice, P. S. and Cohen, B. J., 1996, A school outbreak of parvovirus B19 infection investigated using salivary antibody assays, *Epidemiol. Infect.* **116:**331–338.
54. Fallone, C. A., Elizov, M., Cleland, P., Thompson, J. A., Wild, G. E., Lough, J., Faria, J., and Barkun, A. N., 1996, Detection of *Helicobacter pylori* infection by saliva IgG testing, *Am J. Gastroenterol.* **91:**1145–1149.
55. Simor, A. E., Lin, E., Saibil, F., Cohen, L., Louie, M., Pearen, S., and Donhoffer, H. A., 1996, Evaluation of enzyme immunoassay for detection of salivary antibody to *Helicobacter pylori*, *J. Clin. Microbiol.* **34:**550–553.
56. Christie, J. M., McNulty, C. A., Shepherd, N. A., and Valori, R. M., 1996, Is saliva serology useful for the diagnosis of *Helicobacter pylori?*, *Gut* **39:**27–30.
57. Luzza, F., Maletta, M., Imeneo, M., Marcheggiano, A., Iannoni, C., Biancone, L., and Pallone, F., 1995, Salivary-specific immunoglobulin G in the diagnosis of *Helicobacter pylori* infection in dyspeptic patients, *Am. J. Gastroenterol.* **90:**1820–1823.
58. Patel, P., Mendall, M. A., Khulusi, S., Molineaux, N., Levy, J., Maxwell, J. D., and Northfield, T. C., 1994, Salivary antibodies to *Helicobacter pylori:* Screening dyspeptic patients before endoscopy, *Lancet* **344:**511–512.
59. Salamotova, S. A., Sukhoroslova, L. I., Shalaberidze, I. A., and Brusenina, N. D., 1993, An immunoenzyme method for determining IgA antibodies to lipopolysaccharide in secretions in the diagnosis of shigellosis, *Zh. Mikrobiol. Epidemiol. Immunobiol.* **(3):**85–89.
60. Zachrisson, G., Lagergård, T., Trollfors, B., and Krantz, I., 1990, Immunoglobulin A antibodies to pertussis toxin and filamentous hemagglutinin in saliva from patients with pertussis, *J. Clin. Microbiol.* **28:**1502–1505.

61. da Silva, M. V., Dias Camargo, E., Vaz, A. J., and Batista, L., 1992, Immunodiagnosis of human leptospirosis using saliva, *Trans. R. Soc. Trop. Med. Hyg.* **86:**560–561.
62. Needham, C. S., Lillywhite, J. E., Beasley, N. M., Didier, J. M., Kihamia, C. M., and Bundy, D. A., 1996, Potential for diagnosis of intestinal nematode infections through antibody detection in saliva, *Trans. R. Soc. Trop. Med. Hyg.* **90:**526–530.
63. Masum, M. A. and Evans, D. A., 1994, Agglutinating anti-leishmanial antibodies in the saliva of kala-azar patients, *Trans. R. Soc. Trop. Med. Hyg.* **88:**660.
64. Hajeer, A. H., Balfour, A. H., Mostratos, A., and Cross, B., 1994, *Toxoplasma gondii:* Detection of antibodies in human saliva and serum, *Parasite Immunol.* **16:**43–50.
65. del Muro, R., Acosta, E., Merino, E., Glender, W., and Ortiz-Ortiz, L., 1990, Diagnosis of intestinal amebiasis using salivary IgA antibody detection, *J. Infect. Dis.* **162:**1360–1364.
66. Rujner, J., Socha, J., Barra, E., Gregorcek, H., Madalinski, K., Wozniewicz, B., and Giera, B., 1996, Serum and salivary antigliadin and serum IgA anti-endomysium antibodies as a screening test for coeliac disease, *Acta Paediatr.* **85:**814–817.
67. Bianucci, G. F., Campana, G., Maddali Bongi, S., Palermo, C., and Castagnoli, A., 1993, Salivary immunoglobulins in the diagnosis of primary Sjögren's syndrome, *Minerva Med.* **84:**161–170.
68. Amor, B. and Kahan, A., 1989, Lacrimal and salivary immunoglobulins in Sjögren's syndrome, *J. Autoimmun.* **2:**509–513.
69. Brandtzaeg, P., 1971, Human secretory immunoglobulins VII. Concentrations of parotid IgA and other secretory proteins in relation to the rate of flow and duration of secretory stimulus, *Arch. Oral Biol.* **16:**1295–1310.
70. Rudney, J. D., Kajander, K. C., and Smith, Q. T., 1985, Correlations between human salivary levels of lysozyme, lactoferrin, salivary peroxidase and secretory immunoglobulin A with different stimulatory states and over time, *Arch. Oral Biol.* **30:**765–771.
71. Bratthall, D. and Widerström, L., 1985, Ups and downs for salivary IgA, *Scand. J. Dent. Res.* **93:**128–134.
72. Kugler, J., Hess, M., and Haake, D., 1992, Secretion of salivary immunoglobulin A in relation to age, saliva flow, mood states, secretion of albumin, cortisol, and catecholamines in saliva, *J. Clin. Immunol.* **12:**45–49.
73. Wilton, J. M. A., Renggli, H. H., and Lehner, T., 1976, The isolation and identification of mononuclear cells from the gingival crevice in man, *J. Periodont. Res.* **11:**262–268.
74. Curby, W. A., 1953, Device for collection of human parotid saliva, *J. Lab. Clin. Med.* **41:**493–496.
75. Schaefer, M. E., Rhodes, M., Prince, S. J., Michalek, S. M., and McGhee, J. R., 1977, A plastic intraoral device for the collection of human paratoid saliva, *J. Dent. Res.* **56:**728–733.

5

Genetic Amplification Techniques for Diagnosing Infectious Diseases

XIAOTIAN ZHENG and DAVID PERSING

1. INTRODUCTION

For more than a century, the process of microbial detection, identification, and characterization have relied on the ability to cultivate and purify organisms in the laboratory. *In vitro* growth of an organism, which can be viewed as a biological amplification process, provides for assessment of morphological, biochemical, and metabolic characteristics. These macroscopic, microscopic, and other chemical/immunologic features are collectively known as phenotypes. Unfortunately, phenotypes may be limited in their ability to resolve differences among closely related organisms and therefore sometimes lead to misleading results. In other cases, the organisms are difficult to grow in culture or lack a means of *in vitro* cultivation altogether. This has led to an increasing reliance in recent years on genotypic methods for detecting and characterizing microorganisms.

Genotypic identification methods are based on the characteristics of the genetic information contained within a pathogen. These methods offer a number of potential advantages over traditional methods: (1) Genotypic identifications

XIAOTIAN ZHENG and DAVID PERSING • Division of Clinical Microbiology, Mayo Clinic Foundation, Rochester, Minnesota 55905.

Rapid Detection of Infectious Agents, edited by Specter *et al.* Plenum Press, New York, 1998.

may be potentially more rapid. The majority of genotypic detection methods rely on nucleic acid amplification techniques, which essentially replace biological amplification by rapid amplification of pathogen-specific nucleic acid sequences *in vitro*. (2) Measuring or detecting an organism may not require growth of the organism in the laboratory and thus may allow identification directly from clinical samples. (3) Genotypes may comprise a far greater number of phylogenetically informative data points than phenotypes. For example, a commonly used biochemical strip test for bacterial identification contains 19 substrates (i.e., data points), whereas determining the small subunit ribosomal RNA (rRNA) sequence may lead to deducing several hundred phylogenetically informative positions for identification.

A variety of nucleic acid amplification approaches have been investigated in recent years. All of these approaches fall into one of three categories, target, probe, or signal amplification. PCR is the most prominent example of this technology and is the form that has enjoyed the greatest development. This technique and other target amplification methods (e.g., NASBA, strand displacement amplification or SDA) allow exponential multiplication of a DNA or RNA sequence, beginning with a few copies and producing as many as one billion copies within three hours. Among probe amplification technologies, the ligase chain reaction and Q-beta replicase-based methods both result in a high degree of efficient amplification of bound probe molecules. Signal amplification, which does not use enzyme-catalyzed amplification, is performed most commonly by using branched DNA assays, which have been used widely to quantitate plasma levels of human immunodeficiency virus (HIV) and hepatitis C virus (HCV). These target or signal amplification approaches offer great future potential because of their high sensitivity and because they have been developed for a wide variety of applications. In this review, we provide a brief overview of the available amplification techniques and describe some of their applications. For a more comprehensive review of these technologies and their applications, several recent review articles can be consulted.[1–5]

2. TARGET AMPLIFICATION

2.1. Polymerase Chain Reaction

Since it was first described more than a decade ago, the polymerase chain reaction (PCR) has been used more widely than any other nucleic acid amplification technique. PCR is a powerful method which uses repeated cycles for amplifying selected nucleic acid sequences. Each cycle of PCR consists of three steps: (1) a DNA denaturation step, in which double stranded target DNA is melted at high temperatures, i.e., two strands are separated from each other; (2) a primer annealing step at a lower temperature in which primers (chemically synthesized DNA bearing the complementary sequences specific for the target gene) anneal to their complementary target sequences; and (3) an extension reaction step, in which DNA polymerase extends the sequences between the primers. At the end of each cycle, which consists of the three steps described, the PCR products are theoretically doubled. The whole procedure is done in a programmable thermal cycler. Generally, 30 to 50 thermal cycles are performed. This results in an exponential increase in the total number of DNA copies synthesized.

The PCR products are separated by agarose gel electrophoresis and visualized by DNA staining. The amplification products are also detected by hybridization to a probe that is homologous to an internal sequence of the amplification region. This hybridization step confirms that the amplification product is specifically the sequence of interest and also improves the detection sensitivity.

Some modifications of the standard PCR procedure have been developed in the past several years. A few of these are listed here:

- Multiplex PCR. Multiplex PCR is a PCR reaction in which two or more sets of primer pairs specific for different targets are introduced in the same amplification reaction. Thus, more than one unique target DNA sequence in a specimen is amplified at the same time. This coamplification of multiple targets is used for different purposes. For diagnostic purposes, multiplex PCR is used for detecting internal controls (to help avoid false negative results due to reaction inhibition) and for detecting multiple pathogens from a single specimen. Quantitative competitive PCR is a variation of multiplex PCR. This technique is used to measure quantitatively the amount of target DNA or RNA in a specimen. Quantitative PCR is discussed in more detail in a later section.
- Nested PCR. In nested amplification, one set of primers is used in the first-round amplification for 15 to 30 cycles. Then the amplification products of the first reaction are subjected to a second round of amplification with another set of primers specific for an internal sequence that is amplified by the first primer pair. Nested PCR has extremely high sensitivity, and the final products are often detected without probe hybridization. In addition, the amplification of the first-round amplification product with the second-set primer pair verifies the specificity of the first-round product. The major disadvantage of the nested-amplification protocol is the possibility of contamination during transfer of the first-round amplification products to a second tube. This is avoided by the use of a recently developed single-tube nested protocol. The test is accomplished by physically separating the two amplification mixtures with a layer of oil or by designing the primer sets with substantially different annealing temperatures.[2,6]
- RNA target amplification. In reverse-transcriptase (RT) PCR, RNA tar-

gets are first converted to cDNA and then amplified by PCR. This technique is extremely important because it makes possible the detection of organisms with RNA genomes (e.g., HCV and influenza virus). In addition, bacterial and fungal pathogens may contain hundreds or thousands of mRNA or rRNA copies derived from one or a few DNA genomic targets. Therefore, detection of RNA targets is theoretically more sensitive compared with direct PCR amplification of genomic DNA. Other potential applications of this technique include analysis of gene expression and monitoring the efficacy of antimicrobial therapy. The newly described thermostable DNA polymerase (Tth pol) has efficient reverse transcription activity and therefore is used to detect RNA targets without a separate RT step.[2,7] The elevated reaction temperature increases the stringency of primer hybridization and avoids the possible RNA secondary structure, so that the reaction is more specific and efficient than previous protocols using avian myeloblastosis virus (AMV) RT.

2.2. Transcription-Based Amplification Systems

Transcription-based amplification systems (TAS) are based on amplification by *in vitro* transcription. Described in 1989 by Kwoh *et al.*,[8] the system includes synthesizing a DNA molecule complementary to the target nucleic acid and *in vitro* transcription with the newly synthesized cDNA as a template. Although this technique is used to amplify either DNA or RNA, most applications have been developed for RNA target amplification. The drawback of the original TAS protocol was that it required heat denaturing the RNA-DNA duplex (intermediate product). The enzymes in the system were destroyed by this process and needed to be replenished at the end of each cycle. Newer modifications use the enzymatic activity of *E. coli* RNase H to replace the heat denaturation process. As a result, no changes in the temperature of incubation are required (this is why it is called an isothermal amplification technique). This process is variously called self-sustaining sequence replication (3SR), nucleic acid sequence-based amplification (NASBA), or transcription-mediated amplification (TMA).[2,9] As shown in Fig. 1, the first step of the process is to synthesize DNA that is complementary to the target nucleic acid sequence with reverse transcriptase. The primer used is designed to contain a bacterial phage T7 polymerase-binding site. The RNase H degrades the target RNA in the RNA-DNA duplex. The second primer (which may or may not have the T7 promoter sequence) in the reaction binds to the newly formed cDNA, and RT extends it so that double-stranded cDNAs are formed. The DNA strands containing the T7 promoter are transcription templates so that T7 RNA polymerase can synthesize a large molar excess of RNA. Then these RNA molecules are used as the substrate for the next cycle. The result of this repetitive reaction process is up to a millionfold amplification. More recent

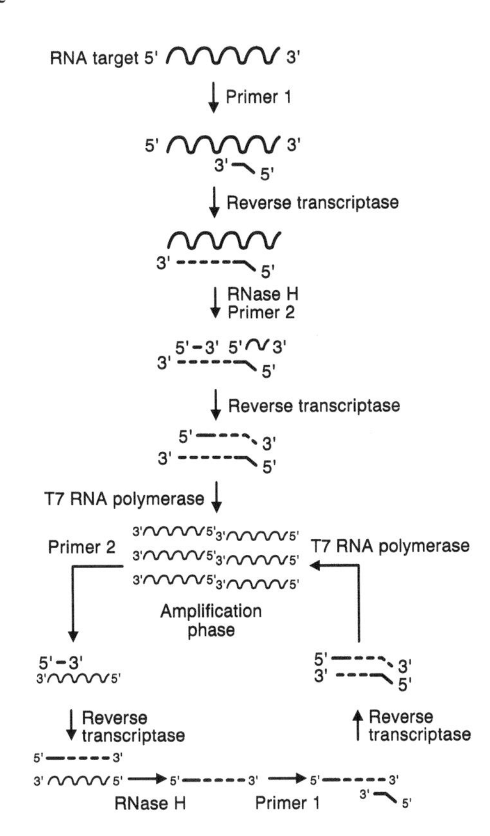

FIGURE 1. Principle of transcription-mediated amplification.

studies demonstrate that because AMV-RT reverse transciptase possesses RNase H activity, the whole reaction is accomplished simply with two enzymes, that is AMV-RT and T7 polymerase.

The detection of the amplification products (RNA) is usually based on specific hybridization with labeled probes. The hybridization is visualized by polyacrylamide gel electrophoresis, or a chemiluminescent signal is read by automated systems.

Although it is technically less robust and less sensitive than PCR, transcription-based amplification has various merits that make it an attractive option: (1) It works at isothermal conditions, so that thermocycling is not required. (2) The test is performed in a single tube to help minimize contamination risks. (3) Amplification of RNA makes it possible to detect viruses, such as HIV and HCV, and also increases the sensitivity of detecting bacterial and fungal pathogens because the inherent biological amplification of RNA targets makes them easier to detect. The first FDA-cleared commercial *Mycobacterium tuberculosis* direct detection test in the U.S. is a TMA-based test that amplifies rRNA as a target (marketed by

GenProbe). Other applications include detecting *Chlamydia trachomatis*, *Mycobacterium avium* complex (MAC), and human papillomavirus.

3. PROBE AMPLIFICATION METHODS

Ligase chain reaction (LCR), also called ligase amplification reaction, is based on target-dependent ligation of two adjacent oligonucleotide probes that hybridize to one strand of the target DNA (Fig. 2). Successful ligation relies on the contiguous positioning and correct base pairing of the 3′ and 5′ ends of oligonucleotide probes on a target DNA molecule, that is, the nucleotide at the 3′ end of the upstream primer needs to coincide with the targeted single base-pair sequence. Differences in this single base pair result in the absence of ligation products. Heating denatures the ligation product and allows the new probes to

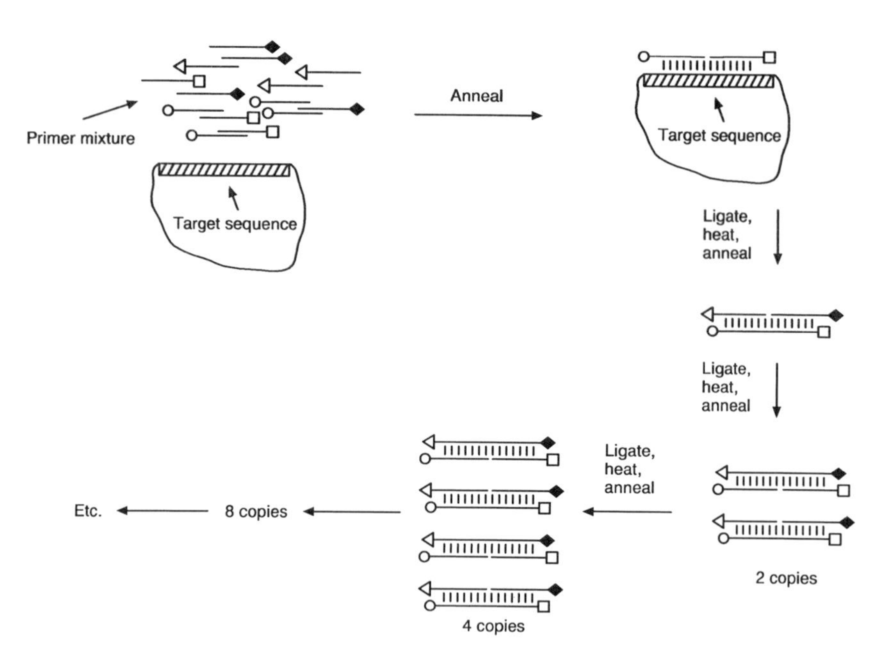

FIGURE 2. LCR. Oligonucleotide probes are annealed head to tail to template molecules with the 3′ end of one probe abutting the 5′ end of the second. DNA ligase joins the adjacent 3′ and 5′ ends to form a duplicate of one strand of the target. Then a second primer set, complementary to the first, uses this duplicated strand (and the original target) as a template for ligation. Repeating the process results in a logarithmic accumulation of ligation products, which can be detected via the functional groups attached to the oligonucleotides. (Reprinted from Ref. 9 with permission from Mayo Foundation, Rochester, Minnesota.)

anneal to the templates at a lower temperature. With a single pair of nucleotide probes, linear amplification is achieved in a cycling reaction. When two pairs of complementary probes are used, exponential amplification is accomplished. Thus, LCR allows discriminating between DNA target sequences that differ in only a single base pair.

The recently developed thermostable DNA ligase greatly simplified this technique and increased its specificity by helping to avoid the problem of blunt-end ligation at low annealing temperature.[10,11] When used following a target amplification method, such as PCR, this technique is sensitive and is useful for detecting point mutations. Newer LCR-based detection methods are convenient and readily automated. If one probe is made with an affinity-labeling molecule, such as biotin, at its 5′ end and the other probe is synthesized with a reporter group such as a fluorescent compound, only ligation products have both the affinity label and reporting molecules. Thus the amplification products can be captured by streptavidin attached on a solid surface like a microtiter plate or beads, so that the products can be detected sensitively.

One potential drawback of LCR is the difficult inactivation of the post-amplification products (Table I). The technique does not allow applying the most widely used contamination control methods. For example, the popular uracil-*N*-glycosylase (UNG) postamplification inactivation protocol cannot be used, because the ligation products are too short to be inactivated by this method. Therefore, effective contamination control in LCR-based techniques is likely to depend to a great extent on physical containment.

TABLE I
Comparison of Contamination Risk Levels of Amplification Technologies

Test		Risk of amplification product contamination[a]
PCR	Commercial product	Low to moderate
	In-house	Moderate to high
SDA	Commercial product	Low to moderate
	In-house	N/A[b]
NASBA	Commercial product	Moderate to high
	In-house	N/A[b]
LCR	Commercial product	Low to moderate
	In-house	N/A[b]
bDNA	Commercial product	Low
	In-house	N/A[b]

[a]Amplification product contamination risk is categorized as low, moderate, and high levels.
[b]N/A = not applicable.

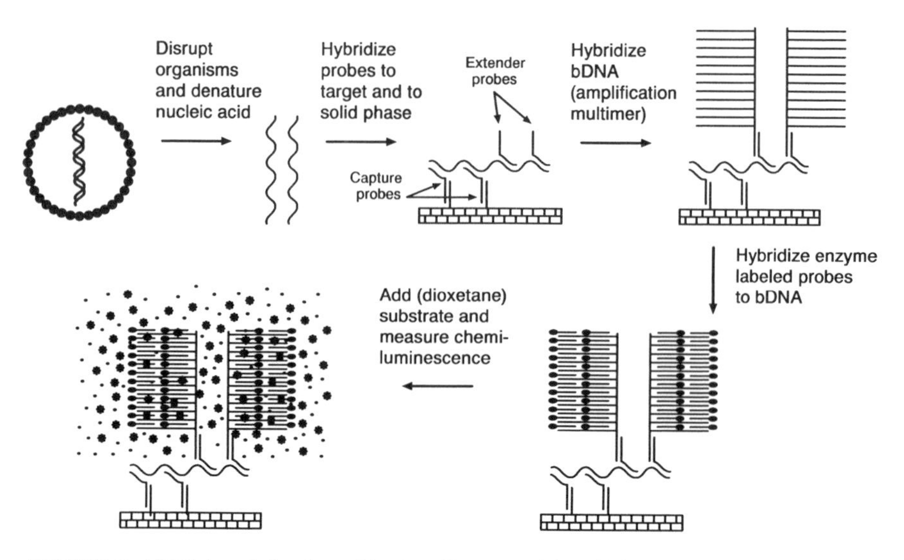

FIGURE 3. bDNA-based signal amplification. Target nucleic acid is released by disruption and captured on a solid surface via multiple contiguous capture probes. Contiguous extender probes hybridize with adjacent target sequences and contain additional sequences homologous to the branched amplification multimer. Enzyme-labeled oligonucleotides bind to the bDNA via homologous base pairing, and the enzyme-probe complex is measured by detecting chemiluminescence. All hybridization reactions occur simultaneously. (Reprinted from Ref. 9 with permission from Mayo Foundation, Rochester, Minnesota.)

4. SIGNAL AMPLIFICATION TECHNIQUES

Signal amplification procedures are based on increasing the signal generated for the probe molecule itself.[2,9] Because signal amplification does not generate a replicable product, it is less likely to have the contamination problems characteristic of target or probe amplification methods. One of its disadvantages is that it is less sensitive and usually detects a minimum of 10^3 to 10^5 nucleic acid targets. Branched DNA (bDNA) technology is a popular example of a signal amplification method. As shown in Fig. 3, it involves several simultaneous hybridization steps. Because bDNA provides multiple discontinuous, target-specific capture and extender probes, it is suitable for detecting nucleic acid target with sequence variations. It is also highly reproducible and thus represents an excellent technological platform for quantitating nucleic acid (see later).

5. QUANTITATIVE NUCLEIC ACID AMPLIFICATION

Although genetic amplification techniques have been studied and widely applied in different fields of biological research and medicine, most tests available in the clinical laboratory provide only qualitative information regarding the status of infectious disease. For many clinical situations, qualitative assessment is sufficient. For instance, information regarding the presence or absence of a particular microorganism (i.e., certain viruses) in a patient may be sufficient to establish a clinical diagnosis. In recent years, however, demand has grown for quantitating nucleic acid targets. This quantitative information will help greatly in diagnoses and prognoses and in therapeutic monitoring.[12]

Until recently it has been very difficult to accomplish quantitative nucleic acid amplification because most of the amplification techniques yield products exponentially until a plateau is reached. Any factor interfering with the exponential nature of the amplification process affects the result of the quantitative assay. Theoretically,[13,14] the accumulation of product in an ideal PCR reaction is approximately expressed by the equation $A_n = A_o (1 + R)^n$, where A_o represents the amount of starting material, A_n the amount of PCR product, n the number of cycles of PCR, and R the efficiency of the PCR reaction whose value is between zero and one. This equation is linearized by a logarithmic transformation and changed into $\log A_n = \log A_o + n \log (1 + R)$. The relationships of these parameters are shown in Fig. 4. With the cycle number n and the amount of accumulated product $\log A_n$, the amount of the original starting template $\log A_o$, given by the y-axis intercept, is deduced [Fig. 4(A)]. In practice however, many factors affect the efficiency of the PCR reaction throughout the amplification procedures, resulting in a difference between the theoretical and actual yield of the reaction. The best results of quantitative PCR analysis are obtained during the exponential phase of the reaction [Fig. 4(B)]. Several reaction parameters affect the performance of quantitative assays, including nucleotide and primer concentrations, accumulation of amplification products, and enzymatic activity. Thus, good controls are critical in quantitative amplification analysis. Indeed, the establishment of appropriate controls is one of the major contributions that has allowed using quantitative amplification more readily in clinical diagnostic laboratories. Such

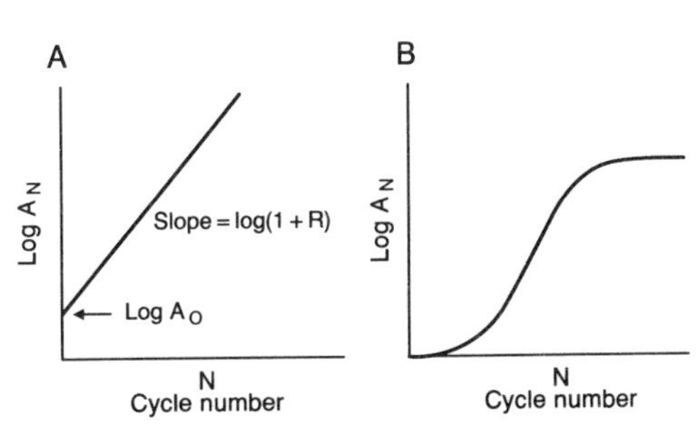

FIGURE 4. (A) Theoretical rate of increase of PCR product as a function of cycle number. When plotted in a log linear fashion, the accumulation of PCR products yields a straight line with a slope equal to $\log(1 + R)$ and a *y*-intercept corresponding to $\log A_o$. See text for details. (B) The plateau effect. In the real world PCR products increase exponentially in the early and middle phases of the reaction, then level off with increasing cycle numbers as substrates are consumed, enzyme is depleted, and increasing copies of PCR products compete with decreasing amounts of primers for primer banding sites. The actual *y*-intercept is often indeterminable. (Reprinted from Ref. 13 with permission from Lippincott-Raven Publishers.)

controls may be internal or external. Internal controls comprise a cellular DNA or RNA target which is coamplified with the target of interest within the same amplification reaction. For quantitative RNA analysis, the internal control is usually an mRNA whose expression is stable under the conditions being studied, such as a "housekeeping" gene. This allows studying relative changes in gene expression. A more widely used control is an external control, which is a known amount of standard DNA or RNA template (usually a plasmid) with the same primer binding sites as the target sequences but with a different internal sequence. This control molecule is amplified in the same reaction together with the target of interest so that the amount of target can be deduced.

More sensitive detection methods are required to detect amplification products in the exponential phase, when the amount of product is usually limited. The recent development of sensitive detection techniques has contributed in part to the success of quantitative amplification assays. Some of these methods include automated detection techniques, such as gel-based detection with an automated sequencer or direct high performance liquid chromatography (HPLC), both of which determine the size of the amplification product very accurately. Sequence-specific detection systems are much better choices however, for clinical diagnosis because the specific hybridization of a probe to the amplification product of interest provides an extra way to ensure high specificity in the test. Different systems have been developed for this purpose. The quantitative HIV test developed by Roche (Amplicor™ HIV Monitor™ Test) uses an immobilized capture probe system. Biotin-labeled primers are incorporated into the product after amplification, and these amplification products are captured by HIV-specific or quantitation standard-specific oligonucleotide probes coated on the microwell plate. After washing, the bound, biotinylated amplicons are quantitatively detected with an avidin-horseradish peroxidase (HRP) conjugate and a colorimetric reaction for HRP. Based on the known input copy number of the quantitation standard RNA, the HIV RNA copy number is calculated. This entire detection process can also be automated. Another very sensitive and reliable way of quantitative detection is an electrochemiluminescense method in which the amplification products are captured on magnetic beads via a biotin-mediated system. Then a sequence-specific oligonucleotide probe labeled with tris (2,2′-bipyridine) ruthenium (II) chelate (TBR) is hybridized to the bead-bound amplification products. When the voltage of the electrode increases to a certain level, oxidation of both tripropylamine and TBR supplied in the reaction occurs and results in light emission at 620 nm. The initial amount of specific amplification products is quantitatively measured by the intensity of this light.

Another method of quantitatively measuring viral RNA uses branched DNA (bDNA) technology. The Quantiplex™ HCV RNA Assay, marketed by Chiron, quantitatively detects RNA in human serum or plasma. As described in Fig. 3, after HCV RNA is captured by target-specific probes which bind to the 5′

untranslated and core regions of the HCV genome, a branched DNA "amplifier" probe and then multiple copies of an alkaline-phosphatase-conjugated detection probe are added to generate the signal. Then the light emission produced by the reaction of the bound alkaline phosphatase with the chemiluminescent substrate is measured quantitatively. The concentration of HCV RNA in specimens is determined by a standard curve obtained by using known concentrations. The entire reaction is done in a microwell plate and is easy to perform. It has been suggested that quantitative HCV RNA detection may be useful in evaluating disease prognosis, identifying candidates for antiviral therapy, and monitoring patients' response to antiviral therapy.[15,16]

The quantitative amplification techniques recently developed are likely to have a great impact on infectious disease practice, especially for viral infections. Quantification of HIV RNA is another example of such an application.[17–20] When used together with other surrogate markers, such as a CD4 cell count, determination of plasma HIV RNA viral load is an early and accurate surrogate marker of disease progression and drug susceptibility. This may result in better predictors of disease progression and outcome and in criteria for initiating and modifying antiviral therapy.

As described previously, despite the recent development of better controls and more sensitive detection systems, the results of quantitative detection still vary greatly because of the nature of the amplification. Therefore, only significant changes in quantity are meaningful at the present time. It has been suggested that only those changes greater than 0.5 log (about threefold) in HIV RNA copies reflect significant differences in viral load in a given patient. Furthermore, only those results obtained from the same laboratory with the same methodology are comparable. Overall, quantitative amplification techniques have been a tremendously successful development in molecular diagnostic microbiology. The systems will be perfected in the near future and have ever wider clinical applications.

6. CONCLUDING REMARKS

Despite the fact that target and signal amplification offer great potential for application in the clinical microbiology laboratory, so far only a few assays have entered the commercial marketplace (Table II). At the time of this review, only four assays based on target amplification have been cleared by the Food and Drug Administration. Now that the "trails have been blazed" by the introduction of these assays, however, many more are expected to be introduced over the next few years. This will create a need for standardized protocols for nucleic acid amplification testing along with the introduction of proficiency testing panels for amplification tests to ensure that the technology lives up to its promises.

TABLE II
Commercially Available Products Using Genetic Amplification Techniques for Detecting Microbial Pathogens

Manufacturer	Product name	Organism detected	Method
Gen-Probe Incorporated	MTD Test	*Mycobacterium tuberculosis*	TMA
	AMP CT	*Chlamydia trachomatis*	TMA
Roche Diagnostic Systems, Inc.	Amplicor™		
	Mycobacterium tuberculosis test	*Mycobacterium tuberculosis*	PCR
	Chlamydia trachomatis test	*Chlamydia trachomatis*	PCR
	C.trachomatis / N.gonorrhoeas test	*C.trachomatis* and *N.gonorrhoeas*	PCR
	HIV-1 test	HIV-1	PCR
	HIV-1 Monitor™ test	HIV-1 quantitative test	Quantitative PCR
	HTLV-I/II test	HTLV-I and HTLV-II	PCR
	Hepatitis C virus test	Hepatitis C virus	PCR
	HCV Monitor™ test	Hepatitis C virus quantitative test	Quantitative PCR
	CMV test	Cytomegalovirus	PCR
	Enterovirus test	Enterovirus	PCR
Abbott Laboratories	LCX® Chlamydia	*Chlamydia trachomatis*	LCR
	LCX® Gonorrhea	*N. gonorrhoeas*	LCR
Chiron	Quantiplex™ HBV DNA assay	Hepatitis B virus quantitative test	Quantitative bDNA
	Quantiplex™ HCV RNA 2.0 assay	Hepatitis C virus quantitative test	Quantitative bDNA
	Quantiplex™ HIV RNA assay	HIV-1	Quantitative bDNA
Organon Teknika	HIV-1 QL	HIV-1	NASBA
	HIV-1 QT	HIV-1 quantitative test	NASBA (quantitative)

REFERENCES

1. Whelen, A. C. and Persing, D. H., 1996, The role of nucleic acid amplification and detection in clinical microbiology, *Annu. Rev. Microbiol.* **50:**349–373.
2. Podzorski, R. P. and Persing, D. H., 1995, Molecular detection and identification of microorganisms, in: *Manual of Clinical Microbiology* (P. R. Murray, E. J. Baron, M. A. Pfaller, F. C. Tenover, and R. H. Yolken, eds.), American Society for Microbiology, Washington, D.C., p. 130.
3. Ronai, Z. and Yakubovskaya, M., 1995, PCR in clinical diagnosis, *J. Clin. Lab. Anal.* **9:**269–283.
4. Taylor, G. R. and Logan, W. P., 1995, The polymerase chain reaction: New variations on an old theme, *Curr. Opinion Biotechnol.* **6:**24–29.
5. Bruce, J. J. 1993/1994, Nucleic acid amplification mediated microbial identification, *Sci. Prog.* **77**(3/4):183–206.
6. Erlich, H. A., Gelfand, D., and Sninsky, J. J., 1991, Recent advances in polymerase chain reaction, *Science.* **252:**1643–1651.
7. Meyers, T. W., and Gelfand, D. H., 1991, Reverse transcription and DNA amplification by a *Thermus themophilus* DNA polymerase, *Biochemistry.* **30:**7661–7666.
8. Kwoh, D. Y., Davis, G. R., Whitefield, K. M., Chappelle, H. L., DiMichele, L. J., and Gingerase, T. R., 1989, Transcription-based amplification systems and detection of amplified human immunodeficiency virus type 1 with a bead-based sandwich hybridization format. *Proc. Natl. Acad. Sci. USA.* **86:**1173–1177.
9. Persing, D. H., Smith, T. F., Tenover, F. C., and White, T. J. (eds.), 1993, *Diagnostic Molecular Microbiology-Principles and Applications,* American Society for Microbiology, Washington, D.C.
10. Barany, F., 1991, Genetic disease detection and DNA amplification using cloned thermostable ligase, *Proc. Natl. Acad. Sci. USA* **88:**189–193.
11. Wiedmann, M., Wilson, W. J., Czajka, J., Luo, J., Barany, F., and Batt, C. A., 1994, Ligase chain reaction (LCR)-overview and applications, *PCR Methods Appl.* **3:**S51–S64.
12. Reischl, U., and Kochanowski, B., 1995, Quantitative PCR: A survey of the present technology, *Mol. Biotechnol.* **3:**55–71.
13. Salomon, R. N., 1995, Introduction to quantitative reverse transcription polymerase chain reaction, *Diagn. Mol. Pathol.* **4:**82–84.
14. Crotty, P. L., Staggs, R. A., Porter, P. T., Killeen, A. A., and McGlennen, R. C., 1994, Quantitative analysis in molecular diagnostics, *Hum. Pathol.* **25:**572–579.
15. Davis, G. L., Lau, J. Y., Urdea, M. S., Neuwald, P. D., Wilber, J. C., Lindsay, K., Perrillo, R. P., and Albrecht, J., 1994, Quantitative detection of hepatitis C virus RNA with a solid-phase signal amplification method: Definition of optimal conditions for specimen collection and clinical application in interferon-treated patients, *Hepatology.* **19**(6):1337–1441.
16. Detner, J., Lagier, R., Flynn, J., Zayati, C., Kolberg, J., Collins, M., Urdea, M., and Sanchez-Pescador, R., 1996, Accurate quantification of HCV RNA from all HCV genotypes using branched DNA (bDNA) technology, *J. Clin. Microbiol.* **34:**901–907.
17. Saksela, K., Stevens, C. E., Rubinstein, P., Taylor, P. E., and Baltimore, D., 1995, HIV-1 messenger RNA in peripheral blood mononuclear cells as an early marker of risk for progression to AIDS, *Ann. Intern. Med.* **123:**641–648.
18. O'Brien, T. R., Blattner, W. A., Waters, D., Eyster, M. E., Hilgartner, M. W., Cohen, A. R., Luban, N., Hatzakis, A., Aledort, L. M., Rosenberg, P. S., Miley, W. J., Kroner, B. L., and Goedert, J. J., 1996, Serum HIV-1 RNA levels and time to development of AIDS in the multicenter hemophilia cohort study, *JAMA,* **276:**105–110.
19. Mellors, J. W., Kingsley, L. A., Rinaldo, C. R., Todd, J. A., Hoo, B. S., Kokka, R. P., and Gupta, P., 1995, Quantitation of HIV-1 RNA in plasma predicts outcome after seroconversion, *Ann. Intern. Med.* **122:**573–579.
20. Mellors, J. M., Rinaldo, C. R., Jr., Gupta, P., White, R. M., Todd, J. A., and Kingsley, L. A., 1996, Prognosis in HIV-1 infection predicted by the quantity of virus in plasma, *Science.* **272:**1167–1170.

6

Ligase Chain Reaction for Detecting Sexually Transmitted Diseases

CHARLOTTE A. GAYDOS and THOMAS C. QUINN

1. INTRODUCTION

Ligase chain reaction (LCR) is an enzymatic method for exponentially amplifying a nucleic acid target sequence that is directed by hapten-conjugated oligonucleotide probes.[1] Four single-stranded probes designed to be complementary to sequences of the target nucleic acid are used so that probes 1 and 3 are complementary to the 3′ and 5′ halves of one strand and probes 2 and 4 are similarly complementary to the other strand.[2] Amplification is accomplished by repeated cycles of heating and cooling to achieve strand separation and specific hybridization by the probes. Each cycle doubles the amplicon upon the ligation of the adjacent probes by the enzyme ligase. Although the sensitivity of LCR is approximately 200–300 target molecules, sensitivity has been improved to fewer than five targets by the development of a modified assay known as gap-LCR, which also requires the use of the thermostable DNA polymerase and the four types of

---

CHARLOTTE A. GAYDOS • Division of Infectious Disease, The Johns Hopkins University School of Medicine, Baltimore, Maryland 21205. THOMAS C. QUINN • Division of Infectious Disease, The Johns Hopkins University School of Medicine, Baltimore, Maryland 21205; and National Institute of Allergy and Infectious Diseases, National Institutes of Health, Bethesda, Maryland 20892-2520.

Rapid Detection of Infectious Agents, edited by Specter *et al.* Plenum Press, New York, 1998.

deoxyribonucleoside triphosphates. LCR does not detect RNA, but a modified assay, called asymmetric gap LCR, that requires reverse transcriptase, detects as little as 20 RNA molecules.[1] LCR is highly specific and has been used to detect microbial agents of infectious diseases, such as *Chlamydia trachomatis*, *Neisseria gonorrhoeae*, *Mycobacterium tuberculosis*, hepatitis B virus, human immunodeficiency virus, and type specific human papilloma virus, and also for detecting nucleic acid sequences found in cancer and genetic diseases.[1,2] Thus far, only LCR assays for *C. trachomatis* and *N. gonorrhoeae* have undergone clinical trials and are approved for clinical use by the Federal Drug Administration (FDA), and these are the focus of this chapter.

2. PRINCIPLE

Clinical samples that have undergone a sample preparation process are added to a unit dose of reaction mixture containing an excess of the four probes, thermostable *Thermus thermophilus* DNA ligase and polymerase, nicotinamide adenine dinucleotide (NAD), magnesium chloride, dCTP, dATP, dGTD, and dTTP in a buffer (pH 7.8).[3] The temperature is raised by the thermocycler to 93 °C to denature or separate the double strands of DNA into single strands. The reaction mixture is cooled to 59 °C to allow hybridization of the four nucleotide probes to complementary single-stranded target sequences of the organism. After hybridization, a gap of three nucleotides is designed between the probes, which is filled with the appropriate nucleotides by taq polymerase. Then ligase covalently joins the pair of probes. This gap of a few nucleotides was designed intentionally to prevent blunt end ligation of the probes. The amplified product is exactly complementary to the original target sequence and subsequently is a target in successive cycles of the amplification reaction. The third segment of the cycle is at 62 °C. Forty cycles are performed by the thermocycler and because DNA target doubles with each cycle, increasing exponentially, an amplification of up to a billionfold can occur. A diagram of a gene target sequence, the position of the probes, and the ligation step is shown in Fig. 1.

The amplicons (amplified DNA) are detected by an automated enzyme immunoassay based on a microparticle method illustrated in Figure 2. All four probes are end-labeled at their nonligation termini with either of two haptens (carbazole and adamantane), so that the amplicons are labeled with both haptens. The amplicons are captured on the microparticles by immobilized antibodies directed at one of the haptens. The other end of the amplicon contains the other hapten and is detected by a second antibody conjugated to alkaline phosphatase. Any unligated probe containing this second hapten is removed by washing before this antibody conjugate is added. Thus, only ligated amplicon is detected, which is present only if the sample contained the original target sequence of the infectious

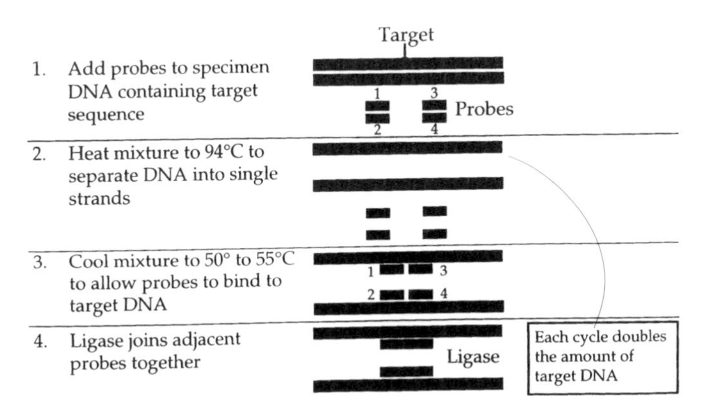

FIGURE 1. The mechanism of ligase chain reaction.

organism. Then the antibody conjugate is detected by adding the substrate 4-methylumbelliferyl phosphate, which is then dephosphorylated by the enzyme to 4-methylumbelliferone, a fluorescent molecule measured optically by the LCX instrument. Negative and positive calibrator controls are used to determine the cutoff value for positive samples.

3. SAMPLE PREPARATION

Preparation of clinical samples varies depending on the organism and the particular sample. The three sample types that are FDA approved for chlamydia and gonorrhea include endocervical swabs, urethral swabs, and urine specimens.

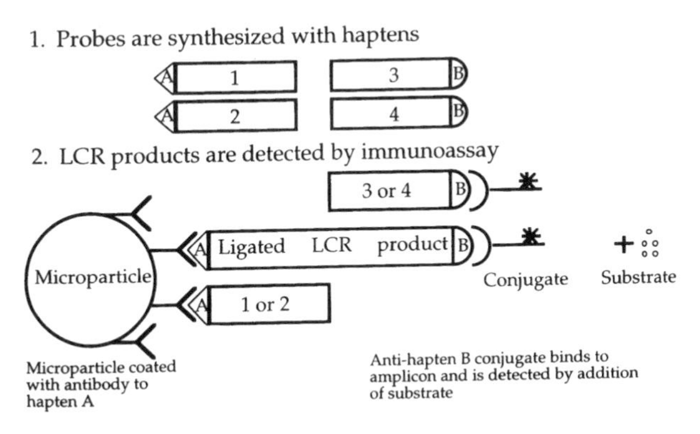

FIGURE 2. Detecting the ligase chain reaction amplicon by microparticle enzyme immunoassay.

Other research studies have demonstrated proficiency with ocular swabs (Quinn, unpublished data) and administered vaginal swabs.[4] All sample preparation techniques must inactivate and lyse the microorganism, while removing biological inhibitors to LCR. Usually, heating to 97 °C (±2 °C) inactivates and also lyses the organism. Some agents require other chemical, enzymatic, or mechanical lysis. Inhibitors are removed by dilution, centrifugation, or both.[1] Ultimately, all LCR assays will have internal controls that will monitor sample inhibitors.[1] Swab specimens in transport tubes supplied by the manufacturer are processed directly, but urine must be centrifuged first at ≥9,000 × g for 15 min. (±2 min.) to produce a pellet for buffer addition and heating.

4. PROBE SET AND TARGET SELECTION

In LCR, the selection of specific gene segments to be amplified depends on the specific organism. Multicopy genes are frequently used to increase sensitivity. For example, for *C. trachomatis,* a 48 base pair sequence of the cryptic plasmid is the target, because there are approximately 10 copies of this plasmid per elementary body.[5] For *N. gonorrhoeae,* a 48-base-pair target of the Opa gene was selected as the target because there are up to 11 copies per cell.[6]

The target sequence is approximately 50 base pairs long allowing each probe member to be long enough for specific hybridization.[1] Targets that have secondary structure are not used, so as not to reduce the efficiency of the LCR. The target sequence selected must be specific to the organism being detected and must not react with other species of the microorganism. This is accomplished by sequence database searches and actually testing related organisms. The chlamydial sequence is highly conserved among all 15 serovars of *C. trachomatis* but is not found in the other species, *C. psittaci* or *C. pneumoniae.*[7] *The target sequence for N. gonorrhoeae* is conserved in all strains thus far studied and is not found in nonpathogenic *Neisseria spp.* or *N. meningitidis.*[8]

5. CONTROL OF CONTAMINATION

Because of the technique's power and like all DNA amplification procedures, LCR is susceptible to contamination and the production of false positive results if samples are contaminated by even minute quantities of amplicon or target DNA. Several methods are used to control possible contamination. Preamplification and postamplification laboratory areas are physically separated and unidirectional traffic and rigorous posttest decontamination of bench and laboratory hood surfaces with bleach and ultraviolet light are used. Aerosol barrier pipet tips for all procedures are employed to prevent cross-contamination. Additionally, single-use, unit doses of preportioned LCR mixtures are provided to prevent contamination during preparation, mixing, or multiuse of reagents. The automated detection of amplicons by the LCX machine eliminates having to open the tubes after amplification because the sampling needle pierces the tube and then destroys the amplicon after detection with a binary inactivating agent capable of reducing DNA by a factor of up to 10^9.[1]

6. LCR APPLICATIONS

6.1. *Chlamydia trachomatis*

C. trachomatis infections are among the most common sexually transmitted diseases among young adults and adolescents.[9,10] It has been estimated that more than 50 million cases occur worldwide and that 4 million cases occur annually in the U.S.[9,10] Infections are associated with many clinical syndromes ranging from urethritis, proctitis, and epididymitis in the male; cervicitis, salpingitis, acute urethral syndrome, endometritis, ectopic pregnancy, infertility, and pelvic inflammatory disease (PID) in the female; and conjunctivitis and pneumonia in infants born to infected mothers.[11–13] Unfortunately, symptoms are often mild or absent among infected men and women, thereby creating a large reservoir of infected persons who continue transmission to new sexual partners.[14] Significant complications and sequelae also pose a considerable economic burden on health care costs. Approximately one million cases of symptomatic PID are diagnosed annually in the U.S. leading to 100,000 surgeries and over 275,000 hospitalizations each year.[15,16] The estimated cost associated with the sequelae of ectopic pregnancy and infertility exceeds $five billion annually.[15–17]

Chlamydial infections occur primarily among young sexually active persons. Prevalence crosses all socioeconomic groups and geographical areas and range from 5–20% in various groups of young adults.[13,18] Because symptoms are absent in approximately half of the infected individuals, these prevalences may be severely underestimated. Thus, widespread screening of individuals at greatest risk, for example, those individuals who are young, sexually active, and have new or multiple partners, has been recommended.[9]

Previously, chlamydia has been detected either by (1) staining chlamydial inclusions grown in tissue culture cells;[19] (2) direct examination of patient material to detect elementary bodies using monoclonal antibodies;[20–22] (3) antigen detection in enzyme immunoassay;[23–26] or (4) nucleic acid hybridization.[27–29] More recently, newer molecular technologies that amplify the DNA of chlamydia in clinical specimens, such as the polymerase chain reaction (PCR)[30–38] and now LCR,[3,39–45] offer expanded sensitivities of detection, while maintaining high specificity.

First described in 1989,[46] LCR was applied to chlamydia first by Dille *et al.*[39] and developed commercially by Abbott Laboratories (Abbott Park, Illinois). The first report of amplifying *C. trachomatis* DNA, which reported probes for both the cryptic plasmid and the OMP-1 gene and compared its sensitivity to PCR, demonstrated that all serovars can be amplified and detected.[39] The sensitivity for both PCR and LCR was reported to be equivalent, and both assays detect DNA from only three elementary bodies. Abbott Laboratories developed a commercial assay utilizing the probes for the plasmid sequence and studied its performance on 4,660 clinical urogenital specimens from multiple study sites (Table I).[47] LCR was compared to culture and an expanded standard protocol that encompassed both culture and an algorithm to resolve discrepant specimens. The method to resolve those specimens discordant between LCR and culture utilized direct fluorescent antibody testing of spun culture sediments and repeat LCR, which targeted OMP-1, a different gene target. The resolved sensitivity and specificity of LCR ranged from 93.5% to 98% and from 99.8% to 100%, respectively.[47] In contrast, the sensitivity of culture ranged from 56.5% to 72.3% (Table I). Twenty-nine patients were culture-positive and LCR-negative. Upon repeat analysis, all of the culture vials were LCR-positive by the discrepant algorithm, indicating that no *C. trachomatis* organisms that had a mutation within the probe sequence, nor were any *C. trachomatis* organisms found that did not contain the cryptic plasmid. Various reports have confirmed the high sensitivity and specificity of the LCR procedure on clinical specimens, including its effectiveness on urine specimens from both males and females.[3,41–45]

The ability to utilize urine as a noninvasive screening specimen to diagnose *C. trachomatis* infections has resulted in a renewed interest in population-based screening for STD. In a multicenter study of men that included 1,585 matched urine specimens for LCR and urethral culture specimens, LCR had a resolved sensitivity of 93.5% to 98% and a specificity of 99.8% to 100%.[42] The sensitivity of urethral culture ranged from 40% to 84.6%, depending on study site. LCR results did not differ between symptomatic and asymptomatic men.

TABLE I
***Chlamydia trachomatis* LCR Assay Performance by Specimen Type**

Specimen type	Number of specimens	Resolved sensitivity		Specificity
		Culture	LCR	LCR
Endocervical	2132	65.0% (152/234)	94.4% (221/234)	99.9% (1896/1898)
Male urethral	542	72.3% (73/101)	98.0% (99/101)	100% (441/441)
Male urine	1043	56.5% (87/154)	93.5% (144/154)	99.8% (887/889)
Female urine	943	59.6% (56/94)	95.7% (90/94)	100% (849/849)

When urine from 1,937 women was evaluated in a multicenter trial, the resolved sensitivity and specificity for urine LCR were 95.7% and 100%, respectively.[44] In contrast, the sensitivity of culture of endocervical swabs was 65%. LCR also performs well in a low-prevalence (3.1%) population and has only slightly decreased sensitivity (87.5%) but excellent specificity (100%).[43] In this population from a family planning clinic, the sensitivity of the enzyme immunoassay of urine was 18.8% and for cervical tissue culture was 56.3%. In another study of urine from 600 STD patients and commercial sex workers from Malaysia, the resolved sensitivity and specificity for LCR for females was 100% and 98.5%, respectively, and 100% and 99.6%, respectively, for males.[45]

6.2. *Neisseria gonorrhoeae*

N. gonorrhoeae, first described by Neisser in 1879, is a gram-negative, nonmotile, nonspore-forming diplococcus, belonging to the family *Neisseriaceae,* which is the etiologic agent of gonorrhea. The other pathogenic species is *N. meningitidis,* to which *N. gonorrhoeae* is genetically closely related. Although *N. meningitidis* is not usually considered a sexually transmitted disease, it may infect the mucus membranes of the anogenital area of homosexual men.[48]

Gonococcal infection is either symptomatic or asymptomatic and causes urethritis, cervicitis, proctitis, Bartholinitis, or conjunctivitis. Gonorrhea is the most frequently reported bacterial infection in the United States. In males, complications include epididymitis, prostatitis, and seminal vesiculitis. In homosexuals, rectal infection and pharyngitis occurs. In females, most cases are symptomatic, and infections of the urethra and rectum often coexist. Complications include PID, pelvic pain, ectopic pregnancy, infertility, FitzHugh Curtis syndrome, chorioamnionitis, spontaneous abortion, premature labor, and infections of the neonate, such as conjunctivitis. Other serious sequelae, such as disseminated gonococcal infection (DGI), occur rarely and result in septicemia, septic arthritis, endocarditis, meningitis, and hemorrhagic skin lesions.[49]

The isolation and identification of *N. gonorrhoeae* on selective media is the currently accepted procedure for diagnosing gonococcal infections.[49] Molecular techniques for identifying, sequencing, and amplifying genes from *N. gonorrhoeae* have obviated the requirement for viable organisms to diagnose infections and for epidemiological typing studies. In particular, the technology to use LCR to amplify DNA from clinical specimens is so powerful that, theoretically, one gene copy in a sample is detectable. The power of this amplification method has also led to the use of unconventional specimens types that are unsuitable for culture, such as urine, vaginal swabs, and tissue from the upper reproductive tract.[50,51]

Birkenmeyer and Armstrong first reported using LCR to detect *N. gonorrhoeae* by testing two probe sets against the Opa genes and one against the pilin gene.[52] The use of four hapten-conjugated probes allowed amplifying DNA from 136 isolates of *N. gonorrhoeae* and none of 124 nongonococcal strains, including *N. meningitidis*. The probe sets were designed for regions of the gonococcal genes, so that gap-filling and ligation happened at sites of mismatch for other neisseria. The short gap formed after hybridization with the adjacent oligonucleotides was filled by DNA polymerase in the presence of dGTP, and the DNA ligase from *Thermus thermophilus* sealed the nick.[52]

When the assay was tested for sensitivity, signals 2.2 to 3.3 times background were generated for as few as 1.1 gonococcal cell equivalents per LCR reaction. At 2.7×10^2 cell equivalents per LCR, signals 21–162 times background were generated, depending on which probe set was used. For specificity assays, none of 124 nongonococcal strains produced signals above background, when tested using 1.3×10^6 cells per assay.[52]

Preliminary testing of 100 genital specimens demonstrated a sensitivity of 100% and a specificity of 97.8%. Bloody and heavily exudative negative specimens did not lose a positive signal when spiked with gonococcal DNA.[52]

A multicenter trial demonstrated that the overall sensitivity and specificity of LCR for *N. gonorrhoeae* from 1539 female endocervical specimens were 97.3% and 99.6%, respectively (Table II).[53] When culture was compared to resolve true positive specimens, the sensitivity and specificity for culture was 83.9% and 100%, respectively. There were three culture-positive specimens not detected by

TABLE II
Performance of the *N. gonorrhoeae* LCR Assay for Detecting *N. gonorrhoeae* in Urogenital Specimens

Assay	Prevalence (%)	Sensitivity (%)	Specificity (%)	PPV (%)[a]	NPV (%)[b]
Cervical LCR[d] (N = 1,539)	7.3	97.3	99.6	95.6	99.8
Cervical cult[d] (N = 1,539)	6.1	83.9	100	NA[c]	NA
Male urethral LCR[d] (N = 808)	32.2	98.5	99.8	99.6	99.3
Male urethral culture[d] (N = 808)	31.1	96.5	100	NA[c]	NA[c]
Female urine[e] (N = 203)	14.3	89.7	98.9	92.9	98.3
Male urine[e] (N = 167)	29.3	98.0	99.2	98.0	99.2

[a]PPV, positive predictive value.
[b]NPV, negative predictive value.
[c]NA, not applicable.
[d]Adapted from Ching *et al.*[53]
[e]Adapted from package insert.

LCR which may have been caused by inhibitors. However, culture missed 18 specimens that were positive by LCR and which were confirmed as true positives by using one of the alternative probe sets developed by Birkenmeyer and Armstrong.[52] The specimens used were from both high-prevalence (15.9%) and low-prevalence (2.7%) populations. The additional detection of positives of LCR over that of culture ranged from 13.5% for STD clinics to 25.9% for obstetrical/gynecology clinics (Lee *et al.*, personal communication). Thus, the advantage of the LCR assay for screening in low-prevalence populations is of great value.

In a multicenter clinical trial, which included both urethral swabs from 808 males and urine from 1639 males, LCR had a sensitivity and specificity of 98.5% and 99.8% for urethral swabs (Table II).[53] LCR demonstrated a 5% extra pick-up of positive specimens compared to culture. For urine specimens that had not been previously frozen, the sensitivities and specificities were as follows: urine from females, 89.7% and 98.9%; from males, 98% and 99.2%, respectively (Table II).

Smith *et al.* used LCR to screen urine for *N. gonorrhoeae* from 283 women attending a STD clinic and compared the results to culture of the endocervix and urethra.[54] Positive LCR results were obtained for 51/54 women who had culture-positive cervical or urethral specimens. Two of 229 women with both cervical and urethral negative cultures had a positive LCR result. Discrepant testing with alternative LCR probe sets revealed that the three urine LCR-negative/endocervix and urethral culture-positive specimens were from truly infected patients, whereas the analysis indicated that the two urine LCR-positive/endocervix and urethral culture-negative specimens were truly positive, also. Thus, the resolved sensitivity, specificity, positive predictive, and negative predictive values for LCR of urine were 94.6%, 100%, 100%, and 98.7%, respectively.[54]

7. SPECIMEN TRANSPORT AND ALTERNATIVE SPECIMEN ADVANTAGES

A great advantage of the LCR method for endocervical swabs is that the specimen in the transport tube is stable at room temperature (or at 4 °C) for up to four days, thus removing the need for stringent transport conditions required for cultures. In addition, the same swab can be used to detect either *C. trachomatis* or *N. gonorrhoeae*. Both of these facts make the LCR test a desirable assay in large screening programs.

Because urine samples are easily obtained and noninvasive, using them to screen for chlamydia and gonorrhea offers a great advantage in large public health screening programs, when there is no opportunity to obtain a cervical or urethral specimen. This is especially true for young sexually active patients, such as those in high school, who may not be in contact with a health clinic.[55] Additionally, because urines can be refrigerated up to four days or frozen for up to six months, there are no stringent transport conditions, as required for culture of endocervical or urethral swabs. A preliminary study of self-administered vaginal swabs indicates that this specimen type also gives highly sensitive and specific results in the LCR assay for detecting chlamydial infections.[4] A swab specimen offers additional advantages in screening programs when it is not feasible to collect urine specimens.

8. SUMMARY

Because commercial applications and kits for the molecular amplicaton of DNA, such as LCR, from genitourinary and other specimens are rapidly becoming available, both research and clinical laboratories must decide whether they can successfully adapt to the molecular identification procedures for *C. trachomatis, N. gonorrhoeae,* and other organisms that cause STDs, as they become available. Manufacturers must also address such issues as specimen preparation, new methods to assess the presence of inhibitors, and the possibility of contamination. As these and other issues are resolved, the amplified technology will move from the research arena to the routine clinical laboratory. Innovative tests, which allow testing for several organisms in a single assay will simplify testing modalities and help drive costs down. Future cost-effective analyses, which consider the costs of the disease sequelae prevented, will also justify the higher costs of amplified DNA testing procedures.

REFERENCES

1. Lee, H. H., and Leckie, G. W., 1995, Infectious disease testing by the ligase chain reaction (LCR), in: *Molecular Biology & Biotechnology a Comprehensive Desk Reference* (H. H. Lee, M. A. Chernesky, J. Schachter, J. D. Burczak, W. W. Andrews, S. Muldoon, G. Leckie, and W. E. Stamm, eds.), VCH, New York, pp. 463–466.
2. Lee, H., 1994, The LCX system—From research to clinical laboratories. *Clin. Chem.* **40:**650.
3. Chernesky, M. A., Jang, D., Lee, H., Burczak, J. D., Hu, H., Sellors, J., Tomazic-Allen, S. J., and Mahony, J. B., 1994, Diagnosis of *Chlamydia trachomatis* infections in men and women by testing first-void urine by ligase chain reaction, *J. Clin. Microbiol.* **32:**2682–2685.
4. Stary, A., Chouieri, B., and Lee, H., 1995, Implications of sensitive molecular diagnosis of *Chlamydia trachomatis* in non-invasive samples types, *Eleventh Meet. Internat. Soc. STD Research* **#041,** p. 43 (Abstract). New Orleans, LA.
5. Palmer, L., and Falkow, S., 1986, A common plasmid of *Chlamydia trachomatis, Plasmid* **16:**52–62.
6. Meyer, T. F., Gibbs, C. P., and Hass, R., 1990, Variation and control of protein expression in Neisseria. *Ann. Rev. Microbiol.* **44:**451–477.
7. Joseph, T., Nano, F. E., and Garon, C. F., 1986, Molecular characterization of *Chlamydia trachomatis* and *Chlamydia psittaci* plasmids, *Infect. Immun.* **51:**699–703.

8. Stern, A., Brown, M., Nicke, P., and Meyer, T. F., 1986, Opacity genes in *Neisseria gonorrhoeae:* Control of phase and antigenic variation, *Cell* **47:**61–71.
9. Centers for Disease Control and Prevention, 1993, Recommendations for the prevention and management of *Chlamydia trachomatis* infections, *MMWR* **42:**1–39.
10. Quinn, T. C., and Cates, W., 1992, Epidemiology of sexually transmitted diseases in the 1990s, in: *Advances in Host Defense Mechanisms* (T. C. Quinn, ed.), Raven Press, New York, (Vol. 8.) pp. 1–37).
11. Cates, W., Rolfs, R. T., and Aral, S. O., 1990, Sexually transmitted diseases, pelvic inflammatory disease, and infertility: An epidemiologic update. *Epidemiol. Rev.* **12:**199–220.
12. Gaydos, C. A., 1993, *Chlamydia trachomatis* infections, *Resid. Staff. Phys. Supplement,* 17–22.
13. Stamm, W. E. and Holmes, K. K., 1990, *Chalmydia trachomatis* infections of the adult, in: *Sexually Transmitted Diseases* (K. K. Holmes, P. A. Mardh, P. F. Sparling, and P. J. Wiesner, eds.), McGraw-Hill, New York, pp. 181–193.
14. Quinn, T. C., Gaydos, C., Shepherd, M., Bobo, L., Hook III, E. W., Viscidi, R., and Rompalo, A., 1996, Epidemiologic and microbiologic correlates of *Chlamydia trachomatis* infection in sexual partnerships, *JAMA* **276:**1737–1742.
15. Centers for Disease Control, 1991, Policy guidelines for the prevention and management of pelvic inflammatory disease (PID), *MM* **40(RR-5):**1–25.
16. Cates, J. W., Rolfs, R. T., and Aral, S. O., 1990, Sexually transmitted diseases, pelvic inflammatory disease, and infertility: an epidemiologic update. *Epidemiol. Rev.* **12:**199–220.
17. Westrom, L., Joesoef, R., Reynolds, G., Hagdu, A., and Thompson, S. E., 1992, Pelvic inflammatory disease and fertility: A cohort study of 1,844 women with laparoscopically verified disease and 657 control women with normal laparoscopic results, *Sex. Transm. Dis.* **19:**185–192.
18. Stamm, W. E., 1988, Diagnosis of *Chlamydia trachomatis genitourinary infections, Ann. Intern. Med.* **108:**710–717.
19. Centers for Disease Control, 1980, Laboratory update: Isolation of *Chlamydia trachomatis* in cell culture, Washington, D.C., U.S. Department Health Human Services.
20. Taylor, H. R., Arawala, N., and Johnson, S. L., 1984, Detection of experimental *Chlamydia trachomatis* eye infections in conjunctival smears and in tissue cultures by use of fluorescein-conjugated antibody, *J. Clin. Microbiol.* **20:**391–395.
21. Uyeda, C. T., Welborn, P., Ellison-Birang, N., Shunk, K., and Tsaouse, B., 1984, Rapid diagnosis of chlamydial infections with Microtrak direct test, *J. Clin. Microbiol.* **20:**948–950.
22. Lidner, L. E., Geerling, S., Nettum, J. A., Miller, S. L., Altman, K. H., and Wechter, S. R., 1986, Identification of chlamydiae in cervical smears by immunofluorescence: Technique, sensitivity, and specificity, *Am. J. Clin. Pathol.* **85:**180–185.
23. Clark, A., Stamm, W. E., Gaydos, C., Welsh, L., Quinn, T. C., Schachter, J., and Moncada, J., 1992, Multicenter evaluation of the antigen chlamydia enzyme immunoassay for diagnosis of *Chlamydia trachomatis* genital infection, *J. Clin. Microbiol* **30:**2762–2764.
24. Gaydos, C., Reichart, C., Long, J., Welsh, L., Neumann, T., Hook, E. W., and Quinn, T. C., 1990, Evaluation of Syva enzyme immunoassay for detection of *Chlamydia trachomatis* in genital specimens, *J. Clin. Microbiol.* **28:**1541–1544.
25. Sanders, J. W., Hook, E. W., Welsh, L. E., Shepherd, M. E., and Quinn, T. C., 1994, Evaluation of an enzyme immunoassay for detection of *Chlamydia trachomatis* in urine of asymptomatic men, *J. Clin. Microbiol.* **32:**24–27.
26. Chan, E. L., Brandt, K., and Horsman, G. G., 1994, A 1-year evaluation of Syva microtrak Chlamydia enzyme immunoassay with selective confirmation by direct fluorescent-antibody assay in a high-volume laboratory, *J. Clin. Microbiol.* **32:**2208–2211.
27. Clarke, L. M., Sierra, M. F., Daidone, B. J., Lopez, N., Covino, J. M., and McCormack, W. M., 1993, Comparison of the Syva MicroTrak enzyme immunoassay and Gen-Probe PACE 2 with cell culture for diagnosis of cervical *Chlamydia trachomatis*infection in a high-prevalence female population, *J. Clin. Microbiol.* **31:**968–971.

28. Warren, R., Dwyer, B., Plackett, M., Pettit, K., Rizvi, N., and Baker, A., 1993, Comparative evaluation of detection assays for *Chlamydia trachomatis*, *J. Clin. Microbiol.* **31:**1663–1666.
29. Stary, A., Teodorowicz, L., Horting-Muller, I., Nerad, S., and Storch, M., 1994, Evaluation of the gen-probe PACE 2 and the MicroTrak enzyme immunoassay for diagnosis of *Chlamydia trachomatis* in urogenital samples, *Sex. Trans. Dis.* **21:**26–30.
30. Jaschek, G., Gaydos, C., Welsh, L., and Quinn, T. C., 1993, Direct detection of *Chlamydia trachomatis* in urine specimens from symptomatic and asymptomatic men by using a rapid polymerase chain reaction assay, *J. Clin. Microbiol.* **31:**1209–1212.
31. Bobo, L., Coutlee, F., Yolken, R. H., Quinn, T., and Viscidi, R. P., 1990, Diagnosis of *Chlamydia trachomatis* cervical infection by detection of amplified DNA with an enzyme immunoassay, *J. Clin. Microbiol.* **28:**1968–1973.
32. Holland, S. M., Gaydos, C. A., and Quinn, T. C., 1990, Detection and differentiation of *Chlamydia trachomatis*, *C. psittaci*, and *C. pneumoniae* by DNA amplification, *J. Infect. Dis.* **162:**984–987.
33. Loeffelholz, M. D., Lewinski, C. A., Silver, S. R., Purohit, A. P., Herman, S. A., Buonagurio, D. A., and Dragon, E. A., 1992, Detection of *Chlamydia trachomatis* in endocervical specimens by polymerase chain reaction, *J. Clin. Microbiol.* **30:**2847–2851.
34. Bauwens, J. E., Clark, A. M., Loeffelholz, M. J., Herman, S. A., and Stamm, W. E., 1993, Diagnosis of *Chlamydia trachomatis* urethritis in men by polymerase chain reaction assay of first-catch urine, *J. Clin. Microbiol.* **31:**3013–3016.
35. Bauwens, J. E., Clark, A. M., and Stamm, W. E., 1993, Diagnosis of *Chlamydia trachomatis* endocervical infections by a commercial polymerase chain reaction assay, *J. Clin. Microbiol.* **31:**3023–3027.
36. Mahony, J. B., Luinstra, K. E., Sellors, J. W., and Chernesky, M. A., 1993, Comparison of plasmid- and chromosome-based polymerase chain reaction assays for detecting *Chlamydia trachomatis* nucleic acids, *J. Clin. Microbiol.* **31:**1753–1758.
37. Bass, C. A., Jungkind, D. L., Silverman, N. S., and Bondi, J. M., 1993, Clinical evaluation of a new polymerase chain reaction assay for detection of *Chlamydia trachomatis* in endocervical specimens, *J. Clin Microbiol.* **31:**2648–2653.
38. Quinn, T. C., Welsh, L., Lentz, A., Crotchfelt, K., Zenilman, J., Newhall, J., and Gaydos, C., 1996, Diagnosis by Amplicor PCR for *Chlamydia trachomatis* infection in urine samples from women and men attending sexually transmitted disease clinics, *J. Clin. Microbiol.* **34:**1401–1406.
39. Dille, B. J., Butzen, C. C., and Birkenmeyer, L. G., 1993, Amplification of *Chlamydia trachomatis* DNA by ligase chain reaction, *J. Clin. Microbiol.* **31:**729–731.
40. Gaydos, C. A., Jang, D., Welsh, L. E., Pare, B., Chernesky, M. A., Sellors, J., Mahony, J., Lee, H., Burczak, J. P., and Quinn, T. C., 1994, Ligase chain reaction (LCR): A novel DNA amplification technique for *Chlamydia trachomatis* (CT) in male urine, *Sex. Transm. Dis.* **21:**S124–S125.
41. Schachter, J., Stamm, W. E., Quinn, T. C., Andrews, W. W., Burczak, J. D., and Lee, H. H., 1994, Ligase chain reaction (LCR) to detect *Chlamydia trachomatis* infection of the cervix, *J. Clin. Microbiol.* **32:**2540–2543.
42. Chernesky, M. A., Lee, H., Schachter, J., Burczak, J., Stamm, W. E., and McCormack, W. M., 1994, Rapid diagnosis of *Chlamydia trachomatis* urethral infection in symptomatic and asymptomatic men by testing first-void urine in a ligase chain reaction assay, *J. Infect. Dis.* **170:**1308–1311.
43. Bassiri, M., Hu, H. Y., Domeika, M. A., Burczak, J., Svensson, L. O., Lee, H. H., and Mardh, P. A., 1995, Detection of *Chlamydia trachomatis* in urine specimens from women by ligase chain reaction, *J. Clin. Microbiol.* **33:**898–900.
44. Lee, H. H., Chernesky, M. A., Schachter, J., Burczak, J. D., Andrews, W. W., Muldoon, S., Leckie, G., and Stamm, W. E., 1995, Diagnosis of *Chlamydia trachomatis* genitourinary infection in women by ligase chain reaction assay of urine, *Lancet* **345:**213–216.
45. Gaydos, C. A., Ngeow, Y. F., Lee, H. H., Canavaggio, M., Welsh, L., Johanson, J., and Quinn,

T. C., 1996, Urine as a diagnostic specimen for the detection of *Chlamydia trachomatis* in Malaysia by ligase chain reaction, *Sex. Transm. Dis.* **23:**402–406.

46. Wu, D. Y. and Wallace, R. B., 1989, The ligase amplification reaction (LAR)—amplification of specific DNA sequences using sequential rounds of template-dependent ligation, *Genomics* **4:**560–569.
47. Burczak, J. D., Chernesky, M. A., Tomazic-Allen, S. J., Quinn, T. C., Carrino, J., Schachter, J., Hu, H., Stamm, W. E., and Lee, H. H., 1994. Application of ligase chain reaction to the detection of *Chlamydia trachomatis* in urogenital specimens from men and women, in: *Chlamydial Infections* (J. Orfila, G. I. Byrne, M. A. Chernesky, J. T. Grayston, R. B. Jones, G. L. Ridgeway, P. Saikku, J. Schachter, W. E. Stamm, and R. S. Stephens, eds.), Societa Editrice Esculapio, Bologna, pp. 322–329.
48. Janda, W. M., Morello, J. A., Lerner, S. A., and Bohnhoff, M., 1983, Characteristics of pathogenic *Neisseria spp.* isolated from homosexual men, *J. Clin. Microbiol.* **17:**85–91.
49. Knapp, J. S. and Rice, R. J., 1995, Neisseria and Branamella, in: *Manual of Clinical Microbiology* (P. R. Murray, E. J. Baron, M. A. Pfaller, F. C. Tenover, and R. H. Yolken, eds.), ASM Press, Washington, D.C., pp. 324–340.
50. Mahony, J. B., Luinstra, K. E., Tyndall, M., Sellors, J. W., Krepel, J., and Chernesky, M., 1996, Multiplex PCR for detection of *Chlamydia trachomatis* and *Neisseria gonorrhoeae* in genitourinary specimens, *J. Clin. Microbiol* **33:**3049–3053.
51. Bevan, C. D., Siddle, N. C., Mumtaz, G., Ridgeway, G. L., Pechertana, S., Pickett, M., Ward, M. E., and Watt, P. J., 1994, *Chlamydia trachomatis* in the genital tract of women with acute salpingitis identified by a quantitative polymerase chain reaction before and after treatment, in: *Chlamydial Infections* (J. Orfila, G. I. Byrne, M. A. Chernesky, J. T. Grayston, R. B. Jones, G. L. Ridgeway, P. Saikku, J. Schachter, W. E. Stamm, and R. S. Stephens, eds.), Societa Editrice Esculapio, Bologna, pp. 350–353.
52. Birkenmeyer, L. and Armstrong, A. S., 1992, Preliminary evaluation of the ligase chain reaction for specific detection of *Neisseria gonorrhoeae*, *J. Clin. Microbiol.* **30:**3089–3094.
53. Ching, S., Lee, H., Hook, E. W. III, Jacobs, M. R., and Zenilman, J., 1995, Ligase chain reaction for detection of *Neisseria gonorrhoeae* in urogenital swab, *J. Clin. Microbiol* **33:**3111–3114.
54. Smith, K. R., Ching, S., Lee, H., Ohhashi, Y., Hu, H. Y., Fisher, H. C. III, and Hook, E. W. III, 1995, Evaluation of ligase chain reaction for use with urine for identification of *Neisseria gonorrhoeae* in females attending a sexually transmitted disease clinic, *J. Clin. Microbiol* **32:**455–457.
55. Gaydos, C. A., Crotchfelt, K. A., Howell, M. R., Kralian, S., Hauptman, P., and Quinn, T. C., 1998, Molecular amplification assays to detect chlamydial infections in urine specimens from high school female students and to monitor the persistence of chlamydial DNA after therapy, *J. Inf. Disease* **177:**417–424.

7

Applications of the Polymerase Chain Reaction

DANNY L. WIEDBRAUK and RICHARD L. HODINKA

1. INTRODUCTION

During the past decade, the proliferation of polymerase chain reaction (PCR) applications has been phenomenal. Few of the biological sciences have failed to benefit from this amazingly versatile methodology. Today, PCR methods are being used by archeologists to identify dead royals; by water science chemists to detect pathogens in water supplies; by food scientists to genotype bacterial and yeast strains and to detect foodborne pathogens; by wildlife biologists to identify game animals confiscated from poachers; and by clinical laboratories to detect previously unrecognized infectious agents. The clinical applications of PCR methods are legion, and no single work can list them all. This chapter summarizes the technology, describes the major clinical applications of PCR, and discusses its limitations.

2. THE POLYMERASE CHAIN REACTION

Developed by researchers at the Cetus Corporation,[1–3] the polymerase chain reaction is an elegantly simple method for the *in vitro* synthesis and ampli-

DANNY L. WIEDBRAUK • Departments of Clinical Pathology and Pediatrics, William Beaumont Hospital, Royal Oak, Michigan 48073. RICHARD L. HODINKA • Departments of Pathology and Pediatrics, Clinical Virology Laboratory, Children's Hospital of Philadelphia and University of Pennsylvania School of Medicine, Philadelphia, Pennsylvania 19104.

Rapid Detection of Infectious Agents, edited by Specter *et al.* Plenum Press, New York, 1998.

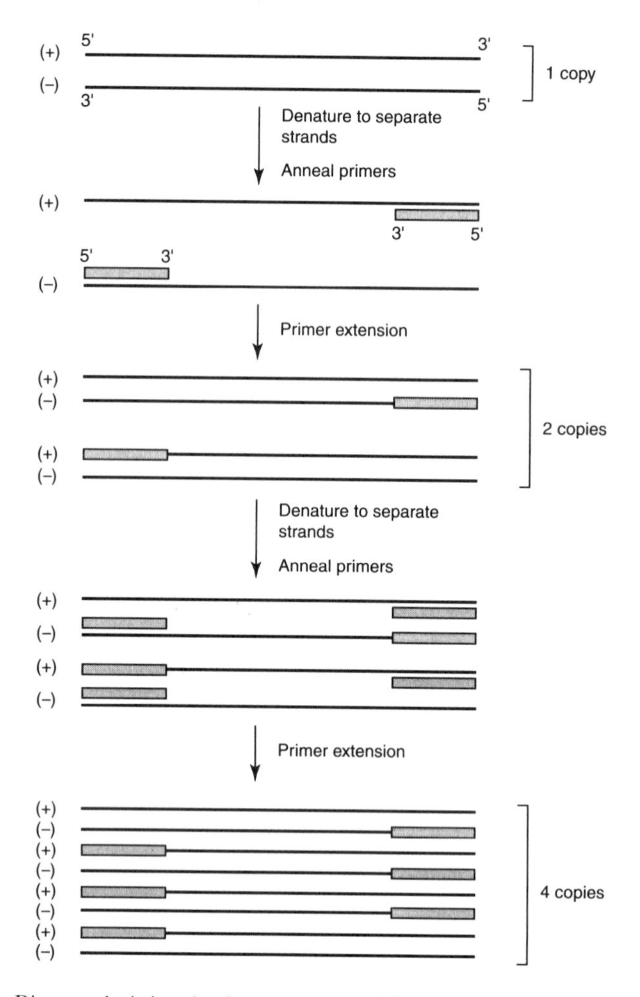

FIGURE 1. Diagram depicting the first two cycles of the polymerase chain reaction. Double-stranded target DNA is separated by heat denaturation. Two synthetic oligonucleotide primers anneal to the separated strands in the 5′ to 3′ orientation and flank the target DNA sequence. *Taq* DNA polymerase initiates new DNA synthesis at the 3′ end of each primer, completing 1 cycle of amplification. After two cycles of synthesis, four copies of the original target DNA molecule have been generated.

fication of specific DNA sequences. To perform PCR, the nucleic acid sequences of an infectious agent must be known, and the target sequences must be unique to that organism (or a group of organisms) to be detected. Two oligonucleotides or primers, typically 15 to 30 bases long, are synthesized so that they are complementary to nucleic acid sequences flanking the region of interest in the target

DNA. In the assay, the primers are added to a reaction mixture containing the target nucleic acid, a heat-stable DNA polymerase, a defined solution of salts, and excess amounts of each of the four deoxynucleoside triphosphates. The mixture is subjected to repeated cycles of temperature changes. These thermal changes facilitate denaturation (94 °C to 97 °C) of the template DNA, annealing (55 °C to 72 °C) of the primers to the target DNA, and extension (72 °C) of the primers so that the target DNA sequence is replicated (Fig. 1). When the reaction mixture is heated again to the denaturation temperature, both the original DNA and the newly synthesized DNA strand serve as templates for another round of DNA replication. Thus, the number of target DNA strands doubles with each cycle. PCR procedures for infectious agents typically include 20 to 40 thermal cycles. A single copy of the target sequence can be amplified 10^5- to 10^6-fold within 3 to 4 hours.

The target nucleic acid may be single- or double-stranded DNA or RNA. For RNA, a reverse transcriptase enzyme must be used to transcribe the RNA into cDNA before amplification. The amplified DNA is detected by one of many methods, including capillary electrophoresis, solid-phase or solution hybridization with appropriate detector probes, high-performance liquid chromatography, and agarose gel electrophoresis with direct visualization of nucleic acids stained with ethidium bromide or comparable reagents.

PCR is both sensitive and specific and is a relatively simple procedure to perform. Because PCR was the first widely used target amplification technology, a large number of clinical applications are available.[4–6]

3. CLINICAL APPLICATION OF THE POLYMERASE CHAIN REACTION

3.1. Early Detection of Infection

Early detection of infectious diseases is extremely important because early treatment often limits the extent of disease, reduces associated sequellae, and improves patient outcome. Early detection of infectious agents can also have a significant financial impact. For instance, early detection of enteroviral meningitis can reduce or eliminate unnecessary hospitalization and antibiotic usage. Early detection of sexually transmitted diseases can reduce the costs associated with pelvic inflammatory disease and the treatment of these infections as they are spread through the community.[7]

However, detecting early infections may be difficult by using traditional assays. Direct fluorescent antibody procedures and enzyme immunoassays (EIA) for detecting microbial antigens often do not have sufficient sensitivity to detect these early infections because the specimens often contain very few organisms or infected cells. Culture methods provide a definitive diagnosis for many infections but culture results are usually not available for days or weeks after the specimens are collected. The enhanced sensitivity of PCR allows the laboratory to detect infectious agents early in the disease process. Because PCR can be completed in 8 to 48 hr, the test information is usually available within a clinically relevant time frame.

Serological testing can also be used to detect early infections. However, serologies are often unrewarding in early disease because agent-specific IgM may not be detectable at the time of presentation. In addition, the IgM response to infectious agents, such as *Toxoplasma gondii,* can persist for up to one year after the primary infection has resolved.[8] Thus, the presence of *T. gondii* IgM may not indicate an active infection. PCR can rapidly identify *T. gondii* in amniotic fluids thereby providing a definitive diagnosis despite confusing serological patterns.

Seroconversion can also be used to document acute infections. For many infectious agents, the "window period" from infection to seroconversion is 7–21 days. This window period can be significantly longer with hepatitis C virus (HCV) where antibody responses may not be detectable until 12 or more weeks after the onset of hepatitis.[9] In contrast to serological tests, PCR methods usually detect HCV viremia 10–19 days after infection—well before the onset of hepatitis.[10]

3.2. Detecting Unculturable Infectious Agents

The use of PCR to detect unculturable infectious agents has significantly changed clinical medicine. Before PCR testing, infections caused by unculturable organisms were inferred on the basis of clinical presentation, tissue histopathology, and/or serological testing. Today, PCR detection of HCV virus and human papillomavirus (HPV) are the gold standards for identifying these agents in clinical specimens. PCR has also made significant contributions to our understanding of the natural history and pathology of these unculturable infectious agents. A representative list of unculturable agents detectable by PCR is shown in Table I.

Many other syndromes exist where an infectious etiology is suspected but the causative agent has not been isolated. The eventual identification of the pathogens associated with sarcoidosis, Kawasaki's disease, and type I diabetes mellitus will most likely be determined through the use of arbitrarily primed PCR and other molecular techniques.[11,12] Detecting and identifying these agents are the first steps toward their eventual control.

3.3. Detecting Slow Growing or Fastidious Agents

Some infectious agents grow slowly *in vitro* and detection using traditional culture methods can require days to weeks. Culture-based tests for slow-growing infectious agents, such as *Mycoplasma pneumoniae* and enteroviruses (Table II), often

TABLE I
Unculturable Human Disease Agents Detected by PCR Methods

Infectious agent	Associated diseases
Astroviruses	Nosocomial and community-acquired gastroenteritis
Bartonella spp.	Bacillary angiomatosis and cat-scratch disease
Hepatitis B virus	Serum hepatitis, associated with hepatocellular carcinoma
Hepatitis C virus	Non-A, non-B hepatitis
Hepatitis D virus	Fulminant hepatitis in patients with hepatitis B coinfections
Hepatitis E virus	Epidemic or enterally transmitted, non-A, non-B hepatitis Associated with waterborne transmission
Hepatitis G virus	Associated with fulminant hepatitis
Human papillomavirus	Oral, skin, and anogenital warts, vaginitis, and cervical carcinoma
Norwalk agents	Sporadic community-acquired gastroenteritis
Whipple's disease bacterium	Systemic disease that usually involves the gastrointestinal tract and the mesentery and also produces intermittent arthralgia

have limited clinical utility because the laboratory results do not contribute to the initial diagnosis or treatment of the patient. In addition, delays associated with *Mycobacterium tuberculosis* cultures can significantly impede disease control efforts.[13] PCR and other amplification methods have significantly improved our ability to detect slow growing agents within a clinically relevant time frame. Rapid identification of infectious agents can also help limit the spread of these agents in the community.

Some infectious agents are fastidious (Table II), and isolating these agents requires special cells or media, special handling, and/or animal inoculation. Detecting these fastidious agents is beyond the capabilities of many hospital laboratories, and the delays in sending specimens to a reference laboratory limits the value of these tests for most practicing physicians. The increasing availability of PCR kits and reagent sets has allowed many hospitals to detect fastidious organisms, such as *T. gondii* and coxsackie A viruses, in a timely fashion and without animal inoculation.

3.4. Detecting Infectious Agents Dangerous to Grow

One of the problems in handling and cultivating infectious agents is the increased risk of acquiring laboratory-associated infections.[14] Laboratory workers are clearly at higher risk for acquiring *Coxiella burnetti, Brucella spp., Francisella tularensis, Chlamydia psittaci,* and certain pathogenic fungi infections than the general population.[15] Although technical and scientific personnel account for most (83.2%) laboratory-acquired infections, janitors, dishwashers, and maintenance

TABLE II
Fastidious or Slow Growing Infectious Agents Detected by PCR

Agent	Associated disease(s)	Problems with isolation
Bartonella henselae	Cat-scratch disease	Does not grow on routine bacteriological media; growth is inhibited by sodium polyanethol-sulfonate (SPS); isolation requires >7 days
BK virus	Urethral stenosis in renal transplant patients and hemorrhagic cystitis in marrow transplant patients	Does not grow in most standard cell cultures; isolation requires weeks to months
Chlamydia pneumoniae	Pneumonia, associated with atherosclerotic heart disease	Cell culture can take 5 days
Coronaviruses	Colds, pneumonia, diarrhea, and necrotizing enterocolitis in newborns	Difficult to grow reproducibly in cell culture
Enteric adenoviruses	Diarrhea	Grows poorly or not at all in standard cell cultures
Enteroviruses	Upper respiratory disease, aseptic meningitis, myocarditis, enteritis, hand, foot, and mouth disease, hemorrhagic conjunctivitis	Not all enteroviruses grow in cell culture; virus isolation requires 2–14 days
Hepatitis A virus	Infectious hepatitis	Isolation methods can take up to 8 weeks
Human parvovirus B19	Erythema infectiosum (fifth disease), hydrops fetalis, chronic anemia, transient aplastic crisis	Does not grow in standard cell cultures; requires bone marrow explant cultures or erythrocyte precursor cultures
JC virus	Progressive multifocal leukoencephalopathy	Requires primary human fetal glial cultures; isolation requires weeks to months
Legionella pneumophila	Legionnaires' disease, community-acquired pneumonia	Fastidious organism; isolation requires 3–5 days
Mycobacterium tuberculosis	Tuberculosis	Slow growth; isolation requires up to 30 days
Mycoplasma pneumoniae	Atypical pneumonia and extrapulmonary manifestations	Culture requires 2–4 weeks
Mycoplasma hominis	Pelvic inflammatory disease, postabortal fever, postpartum fever, pylonephritis, involuntary infertility	Difficult to isolate; isolation procedures can take one week

TABLE II *(Continued)*

Agent	Associated disease(s)	Problems with isolation
Rubella virus	Rubella	Traditional isolation methods can take 2–6 weeks
Toxoplasma gondii	Toxoplasmosis	Isolation requires mouse inoculation; primary isolates rarely recovered in cell cultures

workers account for 13.7% and clerical workers account for 3.7% of infections in one study.[16]

PCR methods reduce the risks of cultivating these infectious agents because the organisms are usually inactivated during the nucleic acid extraction process and because the PCR products are normally not infectious. Although laboratory workers must still handle potentially infectious specimens, PCR can reduce the cutaneous, percutaneous, and aerosol exposures in handling infectious cultures. PCR also allows laboratories without biosafety level (BSL) 3 and 4 facilities to detect BSL-3 and BSL-4 agents in clinical (BSL-2) laboratories.

3.5. Detecting Nonviable Agents

The clinical laboratory is sometimes asked to detect infectious agents late in the course of disease or after the initiation of antimicrobial or antiviral therapies. Conventional culture methods usually produce negative results in these patients because most of the infectious agents are nonviable. The ability of PCR to detect a nonviable virus assists in diagnosing infection when other methods cannot. PCR also detects pathogens in dried blood spots, forensic specimens, and environmental samples where few, if any, viable organisms are present. Detecting infectious agents in immune complexes can help to elucidate the natural history of these infections and has improved our understanding of the immune response to infection.

The ability to detect nonviable infectious agents also presents problems in clinical medicine. Infectious agents killed by appropriate antibiotic usage may still be present at the sampling site and produce a positive result despite effective treatment. Therefore, PCR should not be used as a test of cure because killed organisms persist at some sites for weeks to months. In addition, Kaul *et al.*[17] found that 11 of 55 (20%) washes from sterile bronchoscopes contained residual human DNA and 2 of 55 (3.6%) bronchoscope washes contained *M. tuberculosis* DNA. These findings indicate that residual DNA persists in sterilized bronchoscopes and may be a source of false-positive PCR results.

3.6. Resolving Indeterminate Serologies

PCR methods are valuable for resolving indeterminate human immunodeficiency virus (HIV) and human T-lymphotrophic virus (HTLV) serologies.[18,19] Depending on the study cited, 2 to 49% of specimens that are positive for HIV antibody by enzyme immunoassay (EIA) may be indeterminate by Western blot.[20] 1 to 5% of these specimens are true positives, and these patients will develop antibodies to the *gag, pol,* and envelope proteins within 3 months.[21] Repeat testing in three months is usually sufficient to resolve these indeterminate western blot results. However, indeterminate Western blots are particularly troublesome during pregnancy because the obstetrician must determine whether to initiate antiretroviral therapy, thereby reducing the risk of transmitting HIV to the child.[22]

Detecting HIV infection of newborns born to HIV-seropositive mothers is difficult and time-consuming by culture methods. Antibody tests for HIV are not useful in these cases because these newborns possess maternal antibodies to HIV that may persist for 15 to 18 months after birth. PCR has proven extremely valuable for the early diagnosis of HIV infection in these children.[23]

3.7. Detecting Organisms Present in Low Numbers

Chlamydia trachomatis is the leading cause of sexually transmitted disease in the United States. More than four million new infections occur each year.[7] Left untreated, *Chlamydiae* ascends the female reproductive tract causing cervicitis, uteritis, and salpingitis. Nearly 50% of all pelvic inflammatory disease infections are caused by chlamydiae and one-fifth of these infections produce long-term complications, such as infertility and increased ectopic pregnancy rates.[7] Detecting and treating *C. trachomatis* infections is a formidable public heath challenge because most chlamydia-infected women and many men are asymptomatic and do not seek medical treatment.[27] Many of these asymptomatic individuals have low-level infections that are missed by traditional culture procedures.[7] PCR is significantly more sensitive than culture[24–26] and has largely replaced culture for screening asymptomatic women.

3.8. Small Specimen Volumes

PCR is especially well suited for detecting infectious agents in pediatric, forensic, ocular, and other specimens where tiny sample volumes are the norm. Small sample volumes severely limit the number of agents detectable by conventional methods. In contrast, multiple PCR procedures can be performed on these specimens. For example, five PCR tests (herpes simplex virus (HSV), cytomegalovirus (CMV), varicella-zoster virus (VZV), Epstein-Barr virus (EBV), and human herpesvirus 6) can be performed on a single 100 microliter intraocular fluid specimen—a specimen volume barely sufficient to inoculate one tube or shell vial culture.[27] Multiplex PCR methods can further extend our ability to detect multiple agents from small volume specimens.

3.9. Predicting Antimicrobial/Antiviral Resistance

PCR-based genotypic assays have been developed for rapidly detecting genetic mutations that confer antimicrobial drug resistance. Specific bacterial or viral genes can be amplified by PCR and directly sequenced to identify alterations in the genome known to be associated with resistance to a given antibiotic or antiviral agent. Such assays have been used to detect specific mutations in the reverse transcriptase and protease genes of HIV-1 following the treatment of patients with antiretroviral drugs.[28,29] Clinical isolates of CMV have been screened by PCR to identify mutations in the UL97 phosphotransferase gene and the UL54 DNA polymerase gene that confer resistance to ganciclovir and/or foscarnet.[30] Genotypic analysis following PCR has also been used to detect multidrug resistant *M. tuberculosis*,[31] methicillin-resistant *Staphylococcus aureus*,[32] penicillinase-producing *Neisseria gonorrhoeae*,[33] and to identify antibacterial genes coding for extended-spectrum β-lactamases,[34] aminoglycoside-modifying enzymes,[35] and erythromycin resistance.[36] For viruses and certain bacteria, using PCR to predict drug resistance offers speed and efficiency over phenotypic antimicrobial susceptibility assays in screening multiple isolates. Genotypic resistance assays can be cumbersome and technically demanding, however, and they are not routinely available in most clinical laboratories. These methods also have a disadvantage in that only known drug-resistant mutations are detected and phenotypic drug susceptibility assays are still required to identify drug-resistant microorganisms with novel resistance mutations.

3.10. Genotyping of Antigenically Identical Organisms

PCR has facilitated detecting and typing antigenically similar organisms and has allowed characterizing genetic variants of a given microbial agent. Such genetic analyses can provide useful information about the epidemiological and pathogenic behavior of microbes; detecting microbial colonization and spread within a given patient population; establishing phylogenetic relationships among organisms; recognizing genetic variants that may be resistant or refractile to antimicrobial drugs; and studying species identification and evolution.

Methods, such as arbitrarily primed PCR, interrepeat PCR, and random amplified polymorphic DNA PCR have been described for detecting genetic variations in a number of medically important bacteria, fungi, and parasites.[37,38] All of these techniques are based on the specific selection of primer annealing sites, which leads to the production of amplified fragments of different sizes that can be visualized as fingerprints by gel electrophoresis. PCR fingerprinting[39] has been used to study species evolution, analyze population genetics, and identify new species of the parasites *Trypanosoma, Leishmania, Naeglaria,* and *Giardia.* This technique has been applied to epidemiological studies and the direct identification and species determination of fungi, such as *Aspergillus, Candida,* and *Histoplasma capsulatum.* Molecular typing and epidemiological analysis have also been performed on a number of medically relevant bacteria, including methicillin-resistant *S. aureus, Legionella pneumophila, Campylobacter spp.,* and *Escherichia coli* strains responsible for foodborne gastroenteritis, and *Helicobacter pylori.* Genetic information obtained about patient colonization and infection, outbreak delineation, and treatment failure may play a major role in managing and controlling infections with these organisms.

Viruses, such as HIV-1, HPV, and hepatitis viruses B (HBV) and HCV, demonstrate considerable nucleic acid sequence variability. The development of PCR-based genotyping assays for distinguishing genetic variants of these agents is clinically important for managing infected patients.[28,29,40–42] Viral genetic variation may affect the overall course of disease with these viruses and the responses of patients to antiviral therapy.

Failure of patients with HCV infection to respond to interferon-α therapy, for instance, is the result of the presence of a particular viral genotype. Infection with either genotype 1a or 1b is associated with a poor response to interferon treatment compared with infection with types 2a, 2b, or 3a. Also, the *in vivo* and *in vitro* resistance of HIV-1 to antiretroviral therapy is caused, in part, by the high spontaneous mutation rate in the viral genome during active replication of the virus and to the accumulation of multiple specific mutations in the coding regions of either the reverse transcriptase or the protease of the virus during prolonged therapy. More than 70 recognized HPV types produce epithelial warts of the skin and mucous membranes, but only a select number of these types cause lesions that progress to malignancy. HPV types 5 and 8 are associated with severe squamous cell carcinomas in patients with the rare skin disorder epidermodysplasia verruciformis, and HPV types 16, 18, 31, and 45 are associated with severe cervical dysplasia and the development of cervical carcinoma. HPV types 6 and 11 are rarely found in invasive cancers, but genetic variants of these types have an increased pathogenic potential which can lead to progressive cervical and respiratory cancers. Mutations in the precore/core gene of HBV in patients with hepatitis have led to the development of precore genetic variants that do not produce hepatitis B e antigen. These variants are thought to be associated with severe chronic active hepatitis or acute fulminant hepatitis. The development of mutations in the pre-S/S region of the surface antigen gene of HBV also results in the production of genetic variants capable of evading the immune response induced by HBV vaccines.

TABLE III
Identification of New Infectious Agents by PCR

Infectious agent	Associated disease
Bartonella spp.	Bacillary angiomatosis and cat-scratch disease
Ehrlichia chaffeensis	Human ehrlichiosis
Hepatitis C virus	Non-A, non-B hepatitis
Hepatitis G virus	Fulminant and chronic hepatitis
Human herpesvirus 8	All forms of Kaposi's sarcoma; AIDS-related body-cavity-based lymphomas; non-Kaposi's sarcoma skin lesions; Castleman's disease
Sin Nombre virus	Hantavirus pulmonary syndrome
Tropheryma whippelli	Whipple's disease—systemic disease that usually involves the gastrointestinal tract and the mesentary

3.11. Identifying New Infectious Agents

PCR enables the rapid detection and identification of previously unknown pathogens that cannot be grown in culture, that are present in too low numbers to be detected by more conventional laboratory methods, or are sufficiently different from previously characterized microbial agents.[11,43,44] The procedure relies on the use of primers to conserved or consensus sequences for initial microbial identification and then uses genus or species-specific primers and/or probes to further identify the organism of interest. In this way, diseases of unknown etiology and new microbial agents responsible for defined clinical syndromes are recognized and characterized. The impact of PCR on the association of newly described viral or bacterial agents with established infectious diseases can be seen in Table III.

In 1990, DNA was extracted from bacillary angiomatosis lesions of AIDS patients and was analyzed using PCR and sequencing of conserved bacterial 16S ribosomal RNA (rRNA) to identify a unique bacterium.[45] Phylogenetic analysis of the amplified sequence showed that the agent is closely related to *Rochalimaea quintana,* the agent of trench fever. The organism was eventually classified as *Rochalimaea henselae* and later reclassified in the genus *Bartonella. B. henselae* is now recognized as a cause of bacillary angiomatosis and peliosis hepatitis in HIV-infected patients, and PCR has been used to help establish this organism as the primary etiologic agent of cat-scratch disease.[46] Because this organism is difficult to culture and detect by standard laboratory methods, PCR and other molecular techniques have played a major role in establishing this agent as the etiology of several important diseases.

Similarly, consensus sequence-based PCR was used to identify an organism associated with Whipple's disease.[47] Although rod-shaped bacilli were identified histologically in Whipple's disease lesions many years ago, the suspected bacterium was never cultured or identified by conventional methods. With the advent of PCR, conserved regions of bacterial 16S rRNA were amplified and sequenced directly from infected tissues, and phylogenetic analysis characterized the agent as the actinomycete, *Tropheryma whippelli.*

PCR amplification and sequencing of bacterial 16S rRNA was used again in 1992 to identify a newly recognized organism from patients with ehrlichiosis.[48] Through phylogenetic analysis of the amplified sequence, it was shown that the organism is related but not identical to *Ehrlichia canis,* the cause of canine ehrlichiosis. Unlike *E. canis,* however, initial attempts to isolate the human agent were unsuccessful. This organism was later named *Ehrlichia chaffeensis.*

An outbreak of unexplained acute pulmonary disease occurred in the southwestern United States in May 1993. Although clinical and epidemiological features suggested an infectious disease, no etiological agent was initially recovered. Cross-reactive antibodies to known hantaviruses were soon identified in sera from infected patients, but the antibody response suggested that the disease was caused by a previously unrecognized hantavirus. Conserved sequences of the G2 region of the M segment of known hantaviruses were used in a reverse transcriptase PCR of RNA extracted from diseased tissues to amplify genomic sequences of a hantavirus similar to Prospect Hill virus.[49] Sequence and phylogenetic analysis of the PCR products led to the discovery of a novel hantavirus called Sin Nombre virus. PCR and nucleotide sequence analysis also helped establish that this hantavirus could be maintained and transmitted by the deer mouse, *Peromyscus maniculatus.*

Unique DNA sequences of human herpesvirus type 8 (HHV-8) were first recognized using PCR and representational difference analysis.[50] The technique involves PCR amplification of small DNA fragments present in diseased tissue but absent from healthy tissue of the same patient. Now HHV-8 DNA has been detected by PCR in AIDS-associated and classic Kaposi's sarcoma and that occurring in HIV-negative homosexual men. PCR has also been used to detect HHV-8 DNA in AIDS-related body-cavity-based lymphomas, non-Kaposi's sarcoma skin lesions of transplant patients, Castleman's disease, peripheral blood lymphocytes of HIV-seropositive individuals with and without Kaposi's sarcoma, and prostate tissue and human semen of healthy immunocompetent individuals. PCR and representational difference analysis has also led to the identification of an HHV-8-infected established body-cavity B-lymphoma cell line and the subsequent development of an *in vitro* culture system for growing and characterizing the virus.[51]

Lastly, the molecular cloning of hepatitis C and G viruses has resulted in establishing sensitive PCR assays which have been used to detect these viruses and validate the importance of these agents as causes of chronic and fulminant non-A, non-B hepatitis.[52,53]

TABLE IV
Uses of Multiplex PCR to Detect Bacterial, Viral, or Parasitic Infections

Bacteria
Distinguish *Mycoplasma genitalium* from *Mycoplasma pneumoniae*
Molecular typing and epidemiological survey of *Clostridium perfringens*
Detection or differentiation of *Salmonella, Shigella,* or *Campylobacter* species in stools
Differentiation of enteropathogenic, enterotoxigenic, enteroinvasive, enterohemorrhagic, and enteroaggregative types of *Escherichia coli*
Simultaneous detection of *Actinobacillus actinomycetemcomitans* and *Porphyromonas gingivalis*
Detection of *Bordetella pertussis* strains in nasopharyngeal specimens
Detection of methicillin-resistant *Staphylococcus* species
Detection of *Chlamydia trachomatis, Neiserria gonorrhoeae, M. genitalium, Ureaplasma urealyticum* from patients with urethritis
Detection of *Hemophilus ducreyii* and *Treponema pallidum* in genital ulcerative disease (can also detect herpes simplex virus type 1 and 2 in the same reaction)
Detection and identification of *Yersinia* species
Rapid detection and differentiation of *Mycobacterium* species
Detection of multidrug resistant strains of *M. tuberculosis*
Detection of *H. influenzae* and *Streptococcus pneumoniae* in blood culture
Molecular typing of *H. influenzae*
Viruses
Detection of high- and low-risk types of human papillomaviruses
Simultaneous amplification of multiple HIV-1 gene sequences
Genotyping of hepatitis B, C, E, and G viruses
Detection and typing of human herpesviruses
Simultaneous amplification of respiratory synctial virus, influenza virus types A and B, and parainfluenza virus types 1, 2, and 3
Differentiation of human T lymphotrophic virus types I and II
Detection and differentiation of HIV-1 and hepatitis C virus from blood
Detection of adenovirus type 12, cytomegalovirus, and herpes simplex virus types from patients with celiac disease
Detection of human herpes virus type 6 and cytomegalovirus
Detection of adenovirus and herpes simplex virus types from eyes
Differentiation between polio and non-polio enteroviruses
Detection of herpes simplex virus types 1 and 2 in genital ulcerative disease (can also detect *H. ducreyii* and *Treponema pallidum* in same reaction)
Parasites
Differentiation of toxoplasmosis from AIDS-related central nervous system lymphoma

3.12. Simultaneously Detecting Multiple Agents

Multiplex PCR has been developed for simultaneously coamplifying two or more distinct nucleic acid targets in a single reaction tube. Unlike uniplex PCR, that detects only one gene target at a time, multiplex PCR is more practical and allows the development of diagnostic PCR panels for detecting multiple microbial pathogens from a single specimen. Multiplex PCR methods are especially useful for differentiating microbial species or strains within a single assay. This method has also been used for gender screening, disease linkage analysis, genetic mapping and disease diagnosis, forensic studies, target quantitation, molecular typing, and epidemiological surveys.[54,55] The growing utility of multiplex PCR in detecting bacterial, viral, or parasitic infections is summarized in Table IV.

3.13. Quantitating Infectious Agents

The development of molecular assays to quantitate the levels of virus in infected patients may prove to be one of the most valuable tools to assess the progression of viral diseases, monitor the efficacy of antiviral therapy, predict treatment failure and the emergence of drug-resistant viruses, and to facilitate our understanding of the natural history and pathogenesis of certain viruses. Molecular methods, such as quantitative competitive PCR, reverse transcriptase-PCR, nucleic acid sequence-based amplification, and branched-chain technology are now available for accurately quantitating viral nucleic acids of HIV-1,[56] CMV,[57] EBV,[58] HBV,[59] and HCV.[60] Quantitative PCR assays include those based on externally amplified standards, limited dilutions of target sequences, or competitive amplification with internal standards. The incorporation of internal standards into quantitative PCR is the most informative with regard to controlling the many aspects of assay variability.

In HIV infections, quantitation of viral load is important because high levels of HIV-1 RNA have been related to vertical transmission of the virus from mother to fetus, faster progression to AIDS and death in both adult and pediatric patients, and the emergence of drug resistance during prolonged antiretroviral therapy.[61] Quantitative PCR studies with CMV have demonstrated that patients with active CMV disease have higher levels of CMV DNA and that a rapid rise in CMV DNA copy number correlates with symptoms and drug failure during treatment.[57] Monitoring the quantitative levels of HBV or HCV in patients with chronic hepatitis can assist in identifying those patients who will derive the most benefit from interferon-α therapy and for recognizing treatment failures and relapsing infections.[40,59,60] Using quantitative PCR it has recently been shown that EBV viral load increases several weeks before the development of post-transplant lymphoproliferative disorder (PTLD) in patients who have undergone solid organ transplantations.[58] Such information may be beneficial for preventing the severe morbidity and mortality associated with PTLD by providing more rapid and appropriate management of these patients.

4. PROBLEMS WITH POLYMERASE CHAIN REACTION TESTING

There are a number of technical and procedural issues that need to be considered when performing PCR. Variables that influence the utility of PCR include the appropriate collection and transport of specimens, the nucleic acid target selected, the selection and design of suitable oligonucleotide primers and probes, optimization of specimen preparation and PCR amplification conditions, and the method chosen to detect the amplified product. No single PCR protocol can be applied to all situations, and great care must be taken in choosing the conditions that are right for the microorganism(s) to be detected.

One of the greatest strengths and a major weakness of PCR is its high sensitivity. The generation of millions of copies of DNA from a template sequence can lead to product carryover and cross-contamination of negative specimens with an amplified target. False-negative results can also occur due to interfering substances found in biological specimens. Appropriate processing of clinical specimens and stringent amplicon control measures can limit these problems.

Although PCR provides an exquisitely sensitive and specific method for detecting infectious agents, PCR will never completely replace other laboratory procedures. One reason for this prediction centers on the type of test requests received by virology and microbiology laboratories. PCR performs very well for agent-specific test requests but PCR is not cost-effective when screening for a large number of infectious agents. The presence of normal flora provides an additional level of complexity for microbiological test requests. In these cases, the microbiology laboratory must determine if there are infectious organisms present, and the laboratory must also determine if the presence of these organisms has medical significance. Such determinations cannot be made by current PCR methods.

Most PCR tests in existence today detect a single organism. Therefore, exclusive use of PCR will miss dual infections unless the laboratory is specifically instructed to look for multiple agents. Respiratory specimens often contain more than one virus, and significantly more specimens contain both bacterial and viral agents. Exclusive use of PCR also presumes that the ordering physician knows exactly which infectious agent(s) may be causing the disease. This is clearly not the case because our laboratories detect HSV from approximately 30% of all specimens sent for VZV detection.

Because of its extreme specificity, PCR may not detect new (or even common) infectious agents present in the community. Unless the physician makes the request or the laboratorian decides to test for a given agent, both new and unusual infectious agents can be missed when using PCR methods.

The extreme sensitivity of PCR also makes it difficult to distinguish active infection from colonization. As a result, some qualitative PCR tests may be too sensitive for routine clinical use. For instance, a positive PCR test for *Enterococcus* in a urine specimen does not necessarily indicate that a woman has enterococcal disease. Many women asymptomatically shed enterococci in their urine and enterococcal disease is usually not suspected until there are 50,000 to 100,000 colony forming units/mL. Likewise, a positive PCR for *Pneumocystis carinii* in an immunocompetent patient with pneumonia does not indicate that the pneumonia is caused by *Pneumocystis* because most healthy adults are colonized by this organism. PCR alone also cannot distinguish viral reactivation from primary disease, especially in infections with herpesviruses. Therefore, results for herpesviruses should be interpreted with caution when PCR is used as the sole method for detecting herpesviral disease.

Cost is a major issue in every laboratory, but cost issues are especially acute in managed care environments and in laboratories that are downsizing to remain competitive. Many of these laboratories simply cannot afford to use PCR. Commercial PCR methods are capital-intensive and require dedicated equipment and laboratory space. In addition, the reagent and material costs for commercial PCR assays are 2–5 times the cost of bacterial cultures. The labor costs for PCR and culture are similar when performing a single test. However, labor costs of culture methods become insignificant when one considers that a trained microbiologist can rule out the presence of hundreds of different microorganisms for $2–$10 in labor costs. In contrast, the labor costs for ruling out the same number of organisms using antibody- or PCR-based tests is well over $3,000.

5. CONCLUSIONS

PCR has revolutionized our ability to detect infectious agents in clinical specimens. During the past decade, a large number of extremely sensitive and specific PCR methods have been developed for detecting these agents. Today, PCR is an integral and necessary component of diagnostic and research laboratories. PCR is especially valuable for detecting unculturable infectious agents, fastidious and slow-growing agents, and infectious agents that are dangerous to amplify biologically through culture methods. Quantitative PCR has improved our ability to monitor therapy, detect the development of drug resistance, and to predict the progression of certain infectious diseases. The availability of commercial kits and the continuous improvement and simplification of the technology and instrumentation have moved PCR from an esoteric methodology to a routine clinical tool.

REFERENCES

1. Saiki, R. K., Scharf, S., Faloona, F., Mullis, K. B., Horn, G. T., Erlich, H. A., and Arnheim, N., 1985, Enzymatic amplification of β-globin genomic sequences and restriction site analysis for diagnosis of sickle cell anemia, *Science* **30:**1350–1354.
2. Mullis, K., Faloona, F., Scharf, S., Saiki, R., Horn, G., and Erlich, H., 1986, Specific enzymatic amplification of DNA *in vitro:* The polymerase chain reaction, *Cold Spring Harbor Symp. Quant. Biol.* **51:**263–273.
3. Mullis, K. B. and Faloona, F. A., 1987, Specific synthesis of DNA *in vitro* via a polymerase-catalyzed chain reaction, *Methods Enzymol.* **155:**335–350.
4. Wagar, E. A., 1996, Direct hybridization and amplification applications for the diagnosis of infectious diseases, *J. Clin. Lab. Anal.* **10:**312–325.

5. Whelen, A. C. and Persing, D. H., 1996, The role of nucleic acid amplification and detection in the clinical microbiology laboratory, *Ann. Rev. Microbiol.* **50:**349–373.
6. Engleberg, N. C., 1994, Molecular methods: Applications for clinical infectious diseases, *Ann. Emer. Med.* **24:**490–502.
7. Centers for Disease Control and Prevention, 1993, Recommendations for the prevention and management of *Chlamydia trachomatis* infections, *Morbid. Mortal. Weekly Rep.* **42**(RR-12)**:**1–39.
8. McCabe, R. E. and Remington, J. S., 1990, *Toxoplasma gondii*, in: *Principles and Practice of Infectious Diseases*, 3rd ed. (G. L. Mandell, R. G. J. Douglas, and J. E. Bennett, eds.), Churchill Livingstone, New York, pp. 2090–2103.
9. Lemon, S. M. and Brown, E. A., 1995, Hepatitis C virus, in: *Principles and Practice of Infectious Diseases*, 4th ed. (G. L. Mandell, J. E. Bennett, and R. Dolan, eds.), Churchill Livingston, New York, pp. 1474–1486.
10. Young, K. K., Resnick, R. M., and Myers, T. W., 1993, Detection of hepatitis C virus RNA by a combined reverse transcription-polymerase chain reaction assay, *J. Clin. Microbiol.* **31:**882–886.
11. Relman, D. A., 1993, The identification of uncultured microbial pathogens, *J. Infect. Dis.* **168:**1–8.
12. Gao, S.-J. and Moore, P. S., 1996, Molecular approaches to the identification of unculturable infectious agents, *Emerging Infect. Dis.* **2:**159–167.
13. Centers for Disease Control and Prevention, 1996, Nucleic acid amplification tests for tuberculosis, *Morbid. Mortal. Weekly Rep.* **45**(43)**:**950–952.
14. Harding, L. and Liberman, D. F., 1995, Epidemiology of laboratory-associated infections, in: *Laboratory Safety: Principles and Practice*, 2nd ed. (D. O. Flemming, J. H. Richardson, J. J. Tulis, and D. Vesley, eds.), American Society for Microbiology Press, Washington, DC, pp. 7–15.
15. Collins, C. H., 1993, *Laboratory-Acquired Infections*, 3rd ed. Butterworth-Heinemann, Boston, MA.
16. Wedum, A. G., 1964, Laboratory safety in research with infectious aerosols, *Public Health Rep.* (U.S.) **79:**619–633.
17. Kaul, K.., Luke, S., McGurn, C., Snowden, N., Monti, C., and Fry, W. A., 1996, Amplification of residual DNA sequences in sterile bronchoscopes leading to false-positive PCR results, *J. Clin. Micrbiol.* **34:**1949–1951.
18. Jackson, J. B., MacDonald, K. L., Cadwell, J., Sullivan, C., Kline, W. E., Hanson, M., Sannerud, K. J., Stramer S. L., Fildes, N. J., Kwok, S. Y., Sninsky, J. J., Bowman, R. J., Polesky, H. F., Balfour, H. H., and Osterholm, M. T., 1990, Absence of HIV infection in blood donors with indeterminate Western blot tests for antibody to HIV-1, *N. Eng. J. Med.* **322:**217–222.
19. Kwok, S., Lipka, J. J., McKinney, N., Kellogg, D. E., Poiesz, B., Foung, S. K. H., and Sninsky, J. J., 1990, Low incidence of HTLV infections in random blood donors with indeterminate Western blot patterns, *Transfusion* **30:**491–494.
20. Schleupner, C. J., 1995, Detection of HIV-1 infection, in: *Principles and Practice of Infectious Diseases*, 4th ed. (G. L. Mandell, J. E. Bennett, and R. Dolan, eds.), Churchill Livingstone, New York, pp. 1253–1267.
21. Busch, M. P., Kleinman, S. H., Williams, A. E., Smith, J. W., Ownby, H. E., Laycock, M. E., Lee, L. L. L., Pau, C.-P., Schreiber, G. B., and the Retroviral Epidemiology Donor Study, 1996, Frequency of human immunodeficiency virus (HIV) infection among contemporary anti-HIV-1 and anti-HIV-1/2 supplemental test-indeterminate blood donors, *Transfusion* **36:**37–44.
22. Connor, E. M., Sperling, R. S., Gelber, R., Kiselev, P., Scott, G. O'Sullivan, M. J., Van Dyke, R., Bey, M., Shearer, W., Jacobson, R. L., Jimenez, E., O'Neill, E., Bazin, B., Delfraissy, J.-F., Culnane, M., Coombs, R., Elkins, M., Moye, J., Stratton, P., and Balsley, J., for the Pediatric AIDS Clinical Trials Group Protocol 076 Study Group, 1994, Reduction of maternal-infant transmission of human immunodeficiency virus type 1 with zidovudine treatment, *N. Eng. J. Med.* **331:**1173–1180.
23. Paul, M. O., Tetali, S., Lesser, M. L., Abrams, E. J., Wang, X. P., Kowalski, R., Bamji, M., Napolitano, B., Gulick, L., Bakshi, S., and Pahwa, S., 1996, Laboratory diagnosis of infection

status in infants perinatally exposed to human immunodeficiency virus type 1, *J. Infect. Dis.* **173:**68–76.

24. Ossewaarde, J. M., Rieffe, M., Rosenberg-Arksa, M., Ossenloppele, P. M., Nawrocki, R. P., and van Loon, A. M., 1992, Development and clinical evaluation of a polymerase chain reaction test for detection of *Chlamydia trachomatis*, *J. Clin. Microbiol.* **30:**2122–2128.
25. Vogels, W. H., van Voorst Vader, P. C., and Schröder, F. P., 1993, *Chlamydia trachomatis* infection in a high-risk population: Comparison of polymerase chain reaction and cell culture for diagnosis and follow-up, *J. Clin. Microbiol.* **31:**1103–1107.
26. Toye, B., Peeling, R. W., Jessamine, P., Claman, P., and Gemmill, I., 1996, Diagnosis of *Chlamydia trachomatis* infections in asymptomatic men and women by PCR assay, *J. Clin. Microbiol.* **34:**1396–1400.
27. Werner, J. C. and Wiedbrauk, D. L., 1994, Polymerase chain reaction for diagnosis of herpetic eye disease, *Lab. Med.* **25:**664–667.
28. Katzenstein, D. A., 1995, Viral phenotype and genotype as markers in clinical trials, *J. Acquired Immune Defic. Syndr.* **10**(Suppl. 2)**:**S25–S34.
29. D'Aquila, R. T., 1994, HIV-1 drug resistance. Molecular pathogenesis and laboratory monitoring, *Clin. Lab. Med.* **14:**393–422.
30. Baldanti, F., Sarasini, A., Silini, E., Barbi, M., Lazzarin, A., Biron, K. K., and Gerna, G., 1995, Four dually resistant human cytomegalorvirus strains from AIDS patients: Single mutations in UL97 and UL54 open reading frames are responsible for ganciclovir- and foscarnet-specific resistance, respectively, *Scand. J. Infect. Dis. Suppl.* **99:**103–104.
31. Zhang, Y. and Young, D., 1994, Molecular genetics of drug resistance in *Mycobacterium tuberculosis*, *J. Antimicrob. Chemother.* **34:**313–319.
32. Murakami, K., Minamide, W., Wada, K., Nakamura, E., Teraoka, H., and Watanabe, S., 1991, Identification of methicillin-resistant strains of staphylococci by polymerase chain reaction, *J. Clin. Microbiol.* **29:**2240–2244.
33. Sanchez-Pescador, R., Stempien, M. S., and Urdea, M. S., 1988, Rapid chemiluminescent nucleic acid assays for detection of TEM-1 β-lactamase-mediated penicillin resistance in *Neisseria gonorrhoeae* and other bacteria, *J. Clin. Microbiol.* **26:**1934–1938.
34. Mabilat, C., Goussard, S., Sougakoff, W., Spencer, R., and Courvalin, P., 1990, Direct sequencing of the amplified structural gene and promoter for the extended-broad-spectrum β-lactamase TEM-9 (RHH-1) of *Klebsiella pneumoniae*, *Plasmid* **23:**27–34.
35. Vliegenthart, J. S., Ketelaar-Van Gaalen, P. A. G., and van de Klundert, J. A. M., 1991, Identification of three genes coding for aminoclycoside-modifying enzymes by means of the polymerase chain reaction, *J. Antimicrob. Chemother.* **25:**759–765.
36. Arthur, M., Molinas, C., Mabilat, C., and Courvalin, P., 1990, Detection of erythromycin resistance by the polymerase chain reaction using primers in conserved regions of *erm* rRNA methylase genes, *Antimicrob. Agents Chemother.* **34:**2024–2026.
37. Welsh, J. and McClelland, M., 1990, Fingerprinting genomes using PCR with arbitrary primers, *Nucl. Acids Res.* **18:**7213–7218.
38. Williams, J. G. K., Kubelik, A. R., Livak, K. J., Rafalski, J. A., and Tingey, S. V., 1990, DNA polymorphisms amplified by arbitrary primers are useful as genetic markers, *Nucleic Acids Res.* **18:**6531–6535.
39. van Belkum, A., 1994, DNA fingerprinting of medically important microorganisms by use of PCR, *Clin. Microbiol. Rev.* **7:**174–184.
40. Smith, D. B., Davidson, F., and Simmonds, P., 1995, Hepatitis C virus variants and the role of genotyping, *J. Hepatol.* **23**(Suppl. 2)**:**26–31.
41. Manos, M. M., Ting, Y., Wright, D. K., Lewis, A. J., Broker, T. R., and Wolinsky, S. M., 1989, Use of polymerase chain reaction amplification for the detection of genital human papillomaviruses, *Cancer Cells* **7:**209–214.

42. Brown, J. L., Carman, W. F., and Thomas, H., 1992, The clinical significance of molecular variation within the hepatitis B virus genome, *Hepatology,* **15:** 144–148.
43. Wilson, K. H., 1994, Detection of culture-resistant bacterial pathogens by amplification and sequencing of ribosomal DNA, *Clin. Infect. Dis.* **18:**958–962.
44. Fredricks, D. N. and Relman, D. A., 1996, Sequence-based identification of microbial pathogens: A reconsideration of Koch's postulates, *Clin. Microbiol. Rev.* **9:**18–33.
45. Relman, D. A., Loutit, J. S., Schmidt, T. M., Falkow, S., and Tompkins, L. S., 1990, The agent of bacillary angiomatosis: An approach to the identification of uncultured pathogens, *N. Engl. J. Med.* **323:**1573–1580.
46. Anderson, B., Sims, K., Regnery, R., Robinson, L., Schmidt, M. J., Goral, S., Hager, C., and Edwards, K., 1994, Detection of *Rochalimaea henselae* DNA in specimens from cat scratch disease patients by PCR, *J. Clin. Microbiol.* **32:**942–948.
47. Relman, D. A., Schmidt, T. M., MacDermott, R. P., and Falkow, S., 1992, Identification of the uncultured bacillus of Whipple's disease, *N. Engl. J. Med.* **327:**293–301.
48. Anderson, B. E., Sumner, J. W., Dawson, J. E., Tzianabos, T., Greene, C. R., Olson, J. G., Fishbein, D. B., Olsen-Rasmussen, M., Holloway, B. P., George, E. H., and Azad, A. F., 1992, Detection of the etiologic agent of human ehrlichiosis by polymerase chain reaction, *J. Clin. Microbiol.* **30:**775–780.
49. Ksiazek, T. G., Peters, C. J., Rollin, P. E., Zaki, S., Nochol, S., Spiropoulou, C., Morzunov, S., Feldmann, H., Sanchez, A., Khan, A. S., Mahy, B. W. J., Wachsmuth, K., and Butler, J. C., 1995, Identification of a new North American hantavirus that causes acute pulmonary insufficiency, *Am. J. Trop. Med. Hyg.* **52:**117–123.
50. Chang, Y., Cesarman, E., Pessin, M. S., Lee, F., Culpepper, J., Knowles, D. M., and More, P. S., 1994, Identification of herpesvirus-like DNA sequences in AIDS-associated Kaposi's sarcoma, *Science* **266:**1865–1869.
51. Moore, P. S., Gao, S.-J., Dominguez, G., Cesarman, E., Lungu, O., Knowles, D. M., Garber, R., Pellett, P. E., McGeoch, D. J., and Chang, Y., 1996, Primary characterization of a herpesvirus agent associated with Kaposi's sarcoma, *J. Virol.* **70:**549–558.
52. Cha, T.-A., Kolberg, J., Irvine, B., Stempien, M., Beall, E., Yano, M., Choo, Q.-L., Houghton, M., Kuo, G., Han, J. H., and Urdea, M. S., 1991, Use of a signature nucleotide sequence of the hepatitis C virus for the detection of viral RNA in human serum and plasma, *J. Clin. Microbiol.* **29:**2528–2534.
53. Linnen, J., Wages, J. Jr., Zhang-Keck, Z. Y., Fry, K. E., Krawczynski, K. Z., Alter, H., Koonin, E., Gallagher, M., Alter, M., Hadziyannis, S., Karaylannis, P., Fung, K., Nakatsuji, Y., Shih, J.W.-K., Young, L., Platak, M. Jr., Hoover, C., Fernandez, J., Chen, S., Zou, J.-C., Morris, T., Hyams, K. C., Ismay, S., Lifson, J. D., Hess, G., Foung, S. K. H., Thomas, H., Bradley, D., Margolis, H., and Kim, J. P., 1996, Molecular cloning and disease association of hepatitis G virus: A transfusion-transmissible agent, *Science* **271:**505–508.
54. Edwards, M. C. and Gibbs, R. A., 1994, Multiplex PCR: Advantages, development, and applications, *PCR Methods Appl.* **8:**S65–S75.
55. Mahony, J. B., 1996, Multiplex polymerase chain reaction for the diagnosis of sexually transmitted diseases, *Clin. Lab. Med.* **16:**61–71.
56. Schuurman, R., Descamps, D., Weverling, G. J., Kaye, S., Tijnagel, J., Williams, I., van Leeuwen, R., Tedder, R., Boucher, C. A. B., Brun-Vezinet, F., and Loveday, C., 1996, Multicenter comparison of three commercial methods for quantitation of human immunodeficiency virus type 1 RNA in plasma, *J. Clin. Microbiol.* **34:**3016–3022.
57. Gerdes, J. C., Spees, E. K., Fitting, K., Hiraki, J., Sheehan, M., Duda, D., Jarvi, T., Roehl, C., and Robertson, A. D., 1993, Prospective study utilizing a quantitative polymerase chain reaction for detection of cytomegalovirus DNA in the blood of renal transplant patients, *Transplant. Proc.* **25:**1411–1413.

58. Riddler, S. A., Breinig, M. C., and McKnight, J. L. C., 1994, Increased levels of circulating Epstein–Barr virus (EBV)-infected lymphocytes and decreased EBV nuclear antigen antibody responses are associated with the development of posttransplant lymphoproliferative disease in solid-organ transplant recipients, *Blood* **84:**972–984.
59. Butterworth, L.-A., Prior, S. L., Buda, P. J., Faoagali, J. L., and Cooksley, W. G. E., 1996, Comparison of four methods for quantitative measurement of hepatitis B viral DNA, *J. Hepatol.* **24:**686–691.
60. Hawkins, A., Davidson, F., and Simmonds, P., 1997, Comparison of plasma virus loads among individuals infected with hepatitis C virus (HCV) genotypes 1, 2, and 3 by quantiplex HCV RNA assay versions 1 and 2, Roche monitor assay, and an in-house limiting dilution method, *J. Clin. Microbiol.* **35:**187–192.
61. Saag, M. S., Holodniy, M., Kuritzkes, D. R., O'Brien, W. A., Coombs, R., Poscher, M. E., Jacobsen, D. M., Shaw, G. M., Richman, D. D., and Volberding, P. A., 1996, HIV viral load markers in clinical practice, *Nature Med.* **2:**625–629.

8

Identifying Novel Bacteria Using a Broad-Range Polymerase Chain Reaction

BURT ANDERSON

1. INTRODUCTION

Most bacterial infectious diseases are clinically diagnosed and confirmed by routine laboratory methods. Many bacterial infections resulting from highly fastidious or difficult to isolate organisms, however, are often not detected by these methods. The procedures required to confirm infection by such fastidious bacteria are often either laborious and time-consuming or nonexistent. Some bacterial pathogens, such as some of the rickettsiae and related organisms and other highly fastidious bacteria, have not yet been or have been isolated only recently in the laboratory. In addition, previously unknown bacteria have recently been shown to cause a variety of different diseases including Whipple's disease, cat-scratch disease, and two forms of human ehrlichiosis in the United States. In each of these cases, an infectious agent has been suspected, for the greater part of the century for Whipple's disease and cat-scratch disease, but the agent responsible has not been identified until recently. Thus, it is conceivable that a small but significant percentage of bacterial infections are actually caused by previously unknown bacteria that would not be detected or have not yet been identified by conventional isolation-based methods.

BURT ANDERSON • College of Medicine, Department of Medical Microbiology and Immunology, University of South Florida, Tampa, Florida 33612-4799.

Rapid Detection of Infectious Agents, edited by Specter *et al.* Plenum Press, New York, 1998.

Disease cases with strong clinical evidence of a bacterial etiology are often not supported by laboratory isolation and identification of the responsible pathogen. Fever of unknown origin, culture-negative endocarditis, and "aseptic" meningitis all represent examples where the etiologic agent eludes identification. Likewise, a role for a bacterial pathogen in certain types of "autoimmune disease," sarcoidosis, Kawasaki's disease, Crohn's disease, AIDS-related dementias, and a number of other disease syndromes have been the source of speculation. However, traditional culture-based laboratory methodologies have failed to support disease causality in these cases. Recently, the isolation-centered approach described in Koch's postulates has been questioned.[1] In the light of the powerful molecular techniques and tools available to the modern microbiologist, the association of bacteria with diseased tissues can be accomplished in the absence of culture. Frederick and Relman have described a modernized version of Koch's postulates that they deem appropriate for the molecular era.[1] One particular method, among a variety of other molecular tools, that has proven successful in detecting and identifying new bacteria is a broad-range polymerase chain reaction (PCR) coupled with DNA sequencing.

2. THEORETICAL BASIS OF BROAD-RANGE PCR

Broad-range (or "universal") PCR describes the use of highly conserved DNA sequences as priming sites for PCR amplification of a given gene from a broad range of organisms. In the bacterial kingdom, the 16S ribosomal RNA (rRNA) gene is the most widely used target for such amplification. However, other conserved genes or sequences may also be used. Heat-shock protein genes, genes for enzymes involved in key metabolic functions may also be used. However, their level of conservation is not as great as that of the 16S rRNA gene. The basis and logic behind the use of the 16S rRNA gene as the target for amplification and the methods used for bacterial identification follow.

2.1. The 16S rRNA Gene Is Highly Conserved

The bacterial ribosome consists of two subunits of 30S (small subunit) and 50S (large subunit). The larger subunit consists of two species of rRNA (5S and 23S) and more than 30 associated proteins. The small subunit consists of more than 20 associated proteins and the 16S rRNA that is approximately 1500 nucleotides long. The exact function of the 16S rRNA in translation has not been described but it has been proposed that the folding and unfolding of the complex secondary structure of this molecule is integral to the translocation of the ribosome along the mRNA. Regardless, because translation is the physical link between the genotype and the phenotype, it has been described as the essence of the cell.[2] Thus, the apparent vital function that the 16S rRNA gene plays in the translational machinery of the small subunit of the ribosome explains its high level of conservation among bacteria and even between bacteria and the small subunit rRNA (18S rRNA) of eukaryotic cells. 16S rRNA gene sequence conservation among the kingdom Eubacteria is >60% even among the most distantly related bacteria.

2.2. The 16S rRNA as a "Molecular Chronometer"

Mutations occur in the 16S rRNA gene at the same rate as in any other gene on the bacterial chromosome. However, they are fixed at a slow and somewhat constant rate. The constancy of function of the 16S rRNA in the small subunit of the bacterial ribosome during translation has been hypothesized to explain the slow rate by which mutations are fixed in this gene. In fact, the 16S rRNA gene has frequently been called a "molecular chronometer" which reflects the evolutionary distance between two bacteria over time.[3] Accordingly, the 16S rRNA gene is often used to determine phylogenetic relationships among bacteria and has also recently been used for bacterial taxonomy. An extensive catalog of bacterial 16S rRNA gene sequences exists on databases (e.g., GenBank), permitting rapid comparison of new 16S rRNA gene sequences with those of known bacteria. In this way previously known bacteria are easily identified by comparing an unknown 16S rRNA gene sequence with those in the database. Conversely, the phylogenetic relationship of new or previously unsequenced bacteria can be determined, and the bacteria that are the most closely related to the known organism can easily be identified. Thus, strain variation observed within the 16S rRNA gene of a given bacterial species is usually minimal compared to the variation seen for two unique species that have diverged over a longer period of time. However, a concrete and quantitative definition of the bacterial species from the sole standpoint of 16S rRNA gene sequence has not yet been proposed and may be problematic for certain genera of bacteria.[4]

2.3. Universal Primers—One Primer Pair Fits All

The 16S rRNA gene sequence consists of highly conserved (or universal) regions, variable regions, and regions of intermediate conservation. Interspersed with the conserved regions are nine variable regions identified by Gray *et al.* (Fig. 1).[5] Some of the universal regions are so highly conserved that minimally degenerate primers can be used to prime amplification of the remaining regions, including the variable regions, of the 16S rRNA from virtually any bacteria. These variable regions are all contained within two PCR products amplified by primer pairs EC9 (5′ AAGGATCCTACCTTGTTACGACTT 3′) and EC10 (5′ AATCTAGATTAGATA CCCTDGTAGTCC 3′, where D = A,T, or G) or

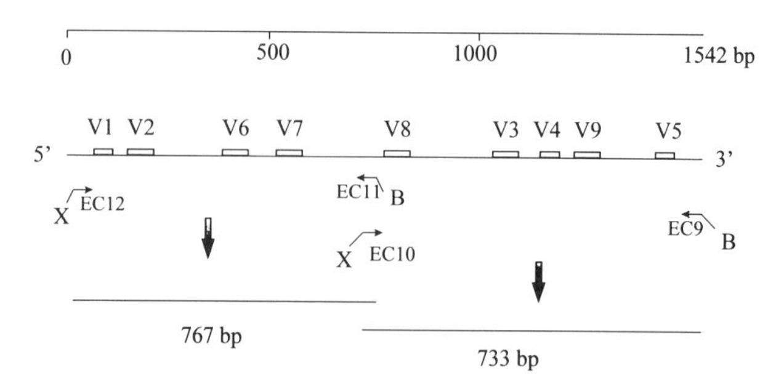

FIGURE 1. Amplification of the 16S rRNA gene (rDNA). The numbering and size of the gene pertain to the gene from *Escherichia coli.* The relative positions of the nine variable regions identified by Gray *et al.*[5] are identified (V1–V9). Modified versions of the primers originally described by Wilson *et al.*[6] are used to amplify 90–95% of the rDNA as two separate PCR products. A product of approximately 767 bp (from *Ehrlichia* spp.) corresponding to the 5′ half of the gene is amplified using primer pair EC11 and EC12. The 3′ half of the gene is amplified as a product of about 733 bp (from *Ehrlichia* spp.) using primer pair EC9 and EC10.

EC11 (5′ AAGGATCCGGACTACHA GGGTATCTAAT 3′ where H = C,T, or A) and EC12 (5′ AATCTAGAGTTTGATCMTGG 3′ where M = A or C). These primers are identical to primers originally described by Wilson *et al.*[6] except that the 5′ end of each has been modified to contain a restriction site to facilitate cloning (*Xba*I site for EC10 and EC12 and *Bam*HI site for EC9 and EC11). Primer pair EC11 and EC12 amplify a product of approximately 767 bp corresponding to the 5′ half of the 16S rRNA gene and EC9 and EC10 amplify a product of 733 bp representing the 3′ half of the gene (Fig. 1). In this way >90% of the 16S rRNA can be amplified as two separate amplicons. In addition to containing highly conserved sequences, these amplicons also contain the variable regions that permit differentiation among more closely related bacteria. One of the first examples of such methodology was by Wilson *et al.*, who used broad-range PCR primers to amplify the 16S rRNA gene from a variety of different bacteria.[6] Similar broad-range primers with slightly different sequences or primers derived from other conserved regions have also been described by other researchers.

2.4. Bacterial Identification without Isolation

Broad-range PCR has evolved in recent years to the point where small numbers of bacteria, such as the amount present in clinical samples, can be amplified by using this technique. Then the resulting PCR-amplified copies of the 16S rRNA gene (often called rDNA) are cloned to facilitate sequencing. The

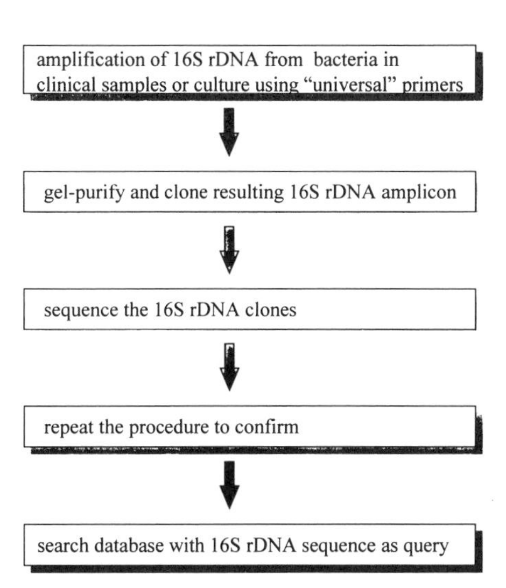

FIGURE 2. General scheme for applying broad-range PCR to identify bacteria. The initial PCR is applied to DNA extracted from bacterial isolates or clinical specimens. Then the resulting PCR products are cloned and sequenced. The procedure is repeated to ensure that the resulting sequence represents the major or sole species in the initial sample and that polymerase incorporation errors are not present (or are identified). Then the sequence obtained is used to query the database of sequences. Bacteria are identified by sequential similarity to existing species (species definition varies with the genus. Generally identical species have >99.0% identity).

clones are sequenced, and the produce is repeated to confirm the sequence (Fig. 2). Alternatively, the amplicon can be sequenced directly. However, direct sequencing is usually not productive in samples with mixed bacterial populations. Once a definitive sequence is obtained, the sequence can be used as a query sequence to search databases (e.g., GenBank) to identify sequences with the greatest level of homology.

Because of the broad range of the universal primers, they amplify DNA from virtually any bacterial contaminant in the laboratory. Although amplicon carry over and laboratory contamination should be a concern in any lab performing PCR, it is especially problematic for broad-range PCR. Precautions to ensure that reagents are not contaminated with bacteria or bacterial DNA should be extensive. Even the Taq polymerase is a known source of contaminating bacterial DNA.[7] Regardless of the precautions observed, it is good practice to confirm the presence and identity of newly described pathogens and their association with disease syndromes by alternative means.

Although this technique is not practical for routine identification of all bacteria, it has proven invaluable in detecting and identifying new or unusual bacterial pathogens. With the advent of automated sequencing equipment, however, the use of this technique has increased and become feasible in both the research and larger clinical laboratory. Examples follow, where broad-range PCR coupled with sequencing have provided powerful tools for identifying bacteria, allowing investigators to solve long-standing chapters in infectious disease mysteries.

3. APPLICATIONS OF BROAD-RANGE PCR

3.1. Ehrlichiosis

In 1987, Maeda *et al.* described a febrile patient with a tick bite history suffering from symptoms similar to those observed with Rocky Mountain spotted fever.[8] Electron microscopy revealed intracellular inclusions characteristic of the rickettsiae belonging to the genus *Ehrlichia.* It was speculated that *E. canis* was the causative agent of this disease based on serological reactivity of patients' sera with *E. canis* antigen.[8] In an attempt to determine if *E. canis* caused human ehrlichiosis, broad-range PCR coupled with sequencing was used to amplify DNA from *Ehrlichia* in blood samples from infected patients. Likewise, the corresponding regions of the 16S rRNA gene were also amplified from blood samples or culture-derived samples of DNA extracted from all known species of *Ehrlichia* (Fig. 3 depicts the 3′ half). However, unique 16S rRNA gene sequences were identified by broad-range PCR amplification of the bacteria in the blood of two patients with ehrlichiosis. These sequences obtained from each of the two patients with ehrlichiosis were identical, but differed from all existing species of *Ehrlichia.*[9] This same sequence was also identified in an isolate subsequently made from one of these two patients.[10] The sequence obtained from both patients differed from all known species of *Ehrlichia* to a greater extent than sequence divergence observed between existing species recognized as distinct (Table I). The name *E. chaffeensis* was proposed for this organism, and the organism has been associated with a number of additional cases of monocytic human ehrlichiosis. PCR remains the only reliable means to identify this rickettsia and differentiate it from the other members of this genus.

3.2. Granulotropic Ehrlichiosis

More recently, the technique of broad-range PCR coupled with sequencing has also been used for resolving the etiology of another form of ehrlichiosis. Three fatal cases occurred in a cluster of patients suffering from ehrlichiosis-like symptoms.[11] Although a number of similarities with ehrlichiosis were observed, several differences were also noted. The initial cases of this new form of ehrlichiosis occurred in Wisconsin and Minnesota, an area outside the known geographic range of *Amblyomma americanum,* the species of tick thought to be responsible for transmission of *E. chaffeensis.*[12] In addition, the new form of ehrlichiosis was observed primarily in granulocytes, a property associated with some species of *Ehrlichia* known to infect animals, but not *E. chaffeensis.* By applying broad-range PCR to DNA extracted from blood samples of patients who had the granulocytic form of ehrlichiosis, it was shown that these patients were infected with another species of *Ehrlichia, E. phagocytophila.*[11] This agent had never been associated with

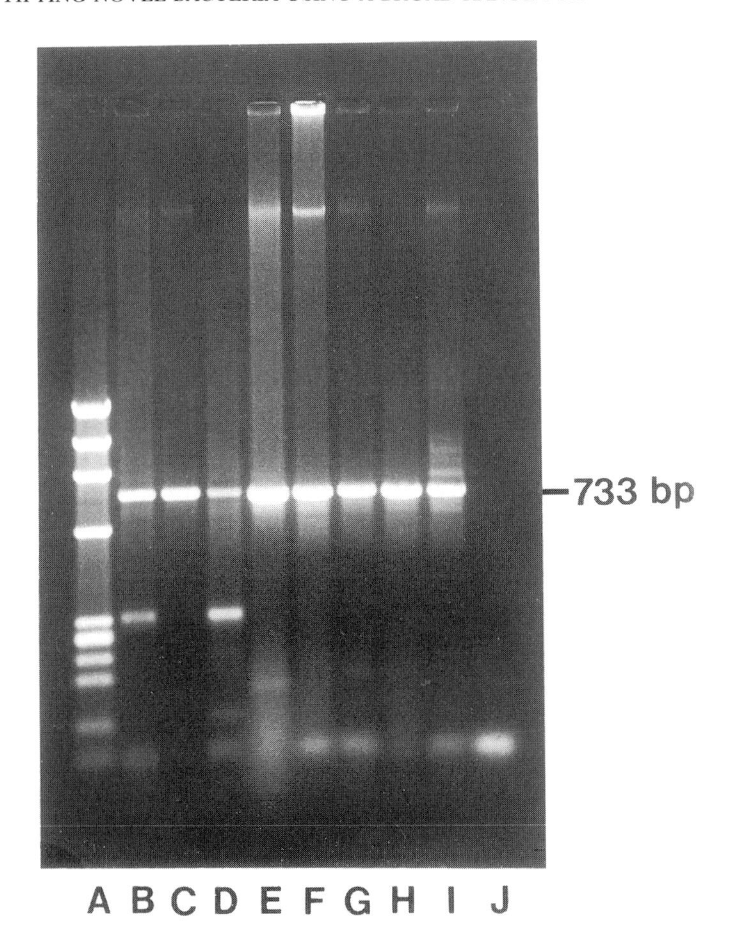

FIGURE 3. Agarose gel electrophoresis of PCR products from different *Ehrlichia* species obtained with primer pair EC9 and EC10. Amplicons include both cultured material and products of direct amplification from the blood of an infected host. The sources of the DNA that amplified are lane B, blood from Arkansas patient; lane C, isolate from Arkansas patient; lane D, blood from Oklahoma patient; lane E, *E. canis* isolate (Oklahoma); lane F, *E. canis* isolate (Florida); lane G, blood from horse infected with *E. equi;* lane H, *E. sennetsu* isolate (Miyayama strain); lane I, *E. sennestu* (11908 strain), and lane J, no DNA control. Lane A contains φX174 phage DNA digested with *Hae*III as molecular size standards (1,353, 1,078, 872, 603, 310, 281/271, 234, 194, 118, and 72 bp, top to bottom). Reproduced with permission. (Anderson *et al.*, 1991).

human disease before and has only recently been isolated and cultivated in the laboratory. The powerful technique of broad-range PCR was used to amplify the *E. phagocytophila* 16S rRNA gene sequences directly from the blood of infected patients. The PCR product was sequenced and identified by comparison with 16S rRNA gene sequences in the GenBank database. Subsequently, the agent of

TABLE I
Relatedness among Ehrlichia 16S rRNA Gene Sequences[a]

Strain	% Sequential relatedness							
	Ech	Eca	Eew	HGE	Eph	Ee	Es	Er
E. chaffeensis (Ech)	–							
E. canis (Eca)	98.1	–						
E. ewingii (Eew)	98.1	98.0	–					
Human granulocytic ehrlichiosis agent (HGE)	92.7	92.3	92.3	–				
E. phagocytophila (Eph)	92.8	92.5	92.4	99.8	–			
E. equi (Ee)	92.7	92.4	92.4	99.8	99.9	–		
E. sennetsu (Es)	84.4	84.2	84.5	85.0	85.1	85.0	–	
E. ristic (Er)	84.2	84.1	83.9	84.8	84.9	84.9	99.0	–

[a]Sequences were aligned pairwise using the "gap" program of the Genetics Computer Group software package.[31]

human granulocytic ehrlichiosis (HGE) has been detected in a number of additional patients and a strong association with *Ixodes dammini* has been established.[13] Thus, it is possible that the same vector responsible for transmitting the agent of Lyme disease (*Borrelia burgdorferi*) may also be responsible for transmitting HGE.

3.3. Whipple's Disease

Whipple's disease is a systemic illness characterized by joint plain, abdominal pain, diarrhea, intestinal malabsorption, and weight loss. In addition, chest pain and chronic nonproductive cough, arthritis, and arthralgia are usually associated with this disease. Diagnosis is made by a jejunal biopsy that upon staining may reveal a bacillus associated with this disease in the affected tissues. Despite numerous attempts to cultivate this organism since the original description of the disease almost 90 years ago,[14] no one has reproducibly cultivated the Whipple's disease bacillus. However, using broad-range PCR coupled with sequencing, Wilson *et al.* were able to identify unique bacterial 16S rRNA gene sequences from a patient with Whipple's disease.[15] Relman *et al.* later confirmed the presence of this bacterial 16S rRNA gene sequence in five additional patients with Whipple's disease.[16] They concluded that the Whipple's disease bacillus is a previously uncharacterized actinomycete. *Tropheryma whippelii* has been proposed as the name of the gram-positive bacillus associated with Whipple's disease.[16] The Whipple's disease bacillus remains an uncultivated bacterium.

3.4. Bacillary Angiomatosis / Cat-Scratch Disease

Bacillary angiomatosis is a disease seen primarily among AIDS patients that present with lesions of the skin and visceral organs. Macroscopically these lesions resemble those observed with Kaposi's sarcoma and are typified by vascular proliferation. Although bacillary angiomatosis was first described in 1983 by Stoler *et al.*,[17] the etiology remained unclear until 1990. Using broad-range PCR primers, Relman *et al.* were able to amplify bacterial 16S rRNA gene sequences from the skin lesions of four patients with bacillary angiomatosis.[18] The sequences from these lesions were identified as belonging to a bacterium related to but not identical to *Rochalimaea quintana*. This organism was subsequently isolated and shown to be a new species, *R. henselae*.[19,20] Reclassification of the genus *Rochalimaea* has resulted in the renaming of all members of this genus to the genus *Bartonella*.[21] Additionally, both *B. henselae* and *B. quintana* have been isolated from the lesions of patients with bacillary angiomatosis and from others with varied disease presentations ranging from relatively mild symptoms to life-threatening bacteremia and endocarditis.[22]

Subsequent to this discovery, *B. henselae* was identified as the primary etiologic agent of cat-scratch disease. Again, broad-range PCR coupled with sequencing also played an important role in this discovery. *B. henselae* 16S rRNA gene sequences were detected in the skin test antigens that had been used for many years to diagnose cat-scratch disease.[23] No other bacterial 16S rRNA gene sequences were detected indicating that the antigenic components of *B. henselae* are likely responsible for eliciting the delayed hypersensitivity observed with the cat-scratch disease skin test. Thus, bacteria that have been causing cat-scratch disease, which was originally described over forty years ago,[24] have avoided detection and identification until the advent of modern molecular techniques.

4. FUTURE DIRECTIONS

4.1. Pathogens versus Normal Flora

The possible applications for using broad-range PCR and sequencing to identify new bacterial pathogens or old pathogens in new places are extensive and varied. In addition the concept of defining bacterial populations in a given environment also warrants study. Defining relative numbers of given bacteria may assist in describing normal flora for various sites, such as the mouth, the vagina, or the intestines. Although this type of study has been done many times before, the nature of culture-based assessments of microbial populations may reflect a bias toward organisms that are more easily cultivated in the laboratory. By creating libraries of 16S rRNA genes belonging to bacteria in a given environment, a sampling can be obtained that is free of the inherent bias present in various isolation and culture techniques. A statistical association between disease state and a given bacterium could be established. For instance, attempts to define the bacterial flora present in healthy women and those with vaginosis may prove fruitful in defining bacterial etiologies not currently associated with vaginosis.

Although this may be labor-intensive, such a project becomes more practical in the era of robotics and automated sequencing.

4.2. Crohn's Disease

In addition a number of disease syndromes that have been the subject of speculation as to a possible bacterial etiology warrant further study. The role of mycobacteria in Crohn's disease has been proposed based on both clinical observations of similarities between intestinal mycobacterioses and this disease. Likewise, specific PCR assays have provided evidence that *Mycobacterium paratuberculosis* is present more often in tissues from patients with Crohn's disease than in controls.[25] However, in that study primers specific for insertion sequences of *Mycobacterium* species were used. Hence, an association between any other bacterial genera and Crohn's disease may have been overlooked. An objective assessment of all bacteria present in diseased tissue might provide insight into the etiology of Crohn's disease.

4.3. Sarcoidosis and Autoimmune Disease

Sarcoidosis is another example of a chronic inflammatory disease that warrants study. Sarcoidosis is characterized by small rounded nonnecrotizing nodules that form on the epithelioid tissue. The etiology of sarcoidosis has been the subject of speculation. Infectious agents on the list of possible causes of sarcoidosis include *Streptococcus* species,[26] *Propionibacterium acnes*,[27] *Mycobacterium* species,[28] and retroviruses,[29] among other agents. Again, sarcoidosis and other chronic inflammatory conditions that might be attributed to "autoimmunity" are fertile grounds to explore a role for a bacterial agent in their etiology. The concept of a persistent bacterial antigen as the catalyst in chronic inflammatory diseases has been the subject of speculation for many years. By applying broad-range PCR to suitable samples (blood, synovial fluid, biopsies) from cases, it might be possible to associate a bacterium with the disease. After the initial association of a given bacterium with a disease syndrome, other more practical means (e.g., serology, antigen detection, specific PCR) could be used to confirm or refute the initial hypothesis of causality.

4.4. Culture-Negative Endocarditis

The incidence of blood culture-negative endocarditis accounts for up to 30% of endocarditis cases. Recently, *Bartonella* species have been identified as a cause of endocarditis. Because this organism is difficult to isolate and requires extended periods of incubation for successful primary isolation, many cases have undoubtedly gone undiagnosed. The optimal conditions for isolation of *Bartonella* species have been described only recently. Raoult *et al.* have recently described 22 new cases of endocarditis from which *Bartonella* species were confirmed by culture, PCR, or serology.[30] The fact that a species of bacteria now recognized as cultivable and identifiable by classical bacteriologic techniques has evaded identification for such a long period of time, provides evidence that other bacterial pathogens may cause "culture-negative" endocarditis. Again, broad-range PCR could provide the means to assess the presence of any bacteria in blood, valvular tissue, or vegetation from patients with endocarditis. In this way it is possible that some cases of culture negative endocarditis might be attributed to a bacterial cause without the bias inherent in culture-based techniques.

5. CONCLUSIONS

The results recently obtained with broad-range PCR coupled with sequencing raise the question of whether other bacterial pathogens exist but remain unidentified because they are not detected by routine clinical microbiological laboratory techniques. Although the technique of broad-range PCR coupled with sequencing is labor-intensive and time-consuming making it impractical for routine clinical laboratories, it is a powerful analytical tool for infectious disease researchers. Further automation of the technique and preparation of clinical specimens will undoubtedly increase the use of broad-range PCR/sequencing and make the technique more practical for the clinical laboratory. Regardless, the use of broad-range PCR and sequencing has already provided new information concerning the etiology of specific bacterial infections. In addition the technique is valuable for researchers interested in the pathology, natural history, and epidemiology of bacterial infectious diseases.

REFERENCES

1. Frederick, D. A. and Relman, D. A., 1996, Sequence-based identification of microbial pathogens: A reconsideration of Koch's postulates, *Clin. Microbiol. Rev.* **9:**18–33.
2. Woese, C., Gutell, R., Gupta, R., and Noller, H., 1983, Detailed analysis of the higher order structure of 16S-like ribosomal ribonucleic acids, *Microbial. Rev.* **47:**621–669.
3. Woese, C., 1987, Bacterial evolution, *Microbiol. Rev.* **51:**221–271.
4. Fox, G. E., Wisotzkey, J. D., and Jurtshuk, P., 1992, How close is close: 16S rRNA sequence identity may not be sufficient to guarantee species identity, *Int. J. Syst. Bacteriol.* **42:**166–170.
5. Gray, M. W., Sankoff, D., and Cedergren, R. J., 1984, On the evolutionary descent of organisms and organelles: A global phylogeny based on a highly conserved structural core in small subunit ribosomal RNA, *Nucleic Acids Res.* **12:**5837–5852.
6. Wilson, K. H., Blitchington, R. B., and Greene, R. C., 1990, Amplification of bacterial 16S ribosomal DNA with polymerase chain reactions, *J. Clin. Microbiol.* **28:**1942–1946.
7. Schmidt, T. M., Pace, B., and Pace, N. R., 1991, Detection of DNA contamination in *Taq* polymerase, *BioTechniques* **11:**176–177.

8. Maeda, K., Markowitz, N., Hawley, R. C., Ristic, M., Cox, D., and McDade, J. E., 1987, Human infection with *Ehrlichia canis*, a leukocytic rickettsia, *N. Engl. J. Med.* **316:**853–856.
9. Anderson, B. E., Dawson, J., Jones, D., and Wilson, K., 1991, *Ehrlichia chaffeensis,* a new species associated with human ehrlichiosis, *J. Clin. Microbiol.* **29:** 2838–2842.
10. Dawson, J., Anderson, B. E., Fishbein, D. B., Sanchez, J. L., Goldsmith, C. S., Wilson, K. H., and Duntley, C. W., 1991, Isolation and characterization of an *Ehrlichia* sp. from a patient diagnosed with human ehrlichiosis, *J. Clin. Microbiol.* **29:**2741–2745.
11. Anderson, B., Sims, K., Olson, J., Childs, J., Piesman, J., Happ, C., Maupin, G., and Johnson, B., 1993, *Amblyomma americanum:* A potential vector of human ehrliciosis, *Am. J. Trop. Med. Hyg.* **49**(2)**:**239–244.
12. Chen, S.-M., Dumler, J. S., Bakken, J. S., and Walker, D. H., 1994, Identification of a granulocytic *Ehrlichia* species as the etiologic agent of human disease, *J. Clin. Microbiol.* **32:**589–595.
13. Pancholi, P., Kolbert, C., Mitchell, P., Reed, K., Dumler, J., Bakken, J., Telford, S., and Persing, D., 1995, *Ixodes dammini* as a potential vector of human granulocytic ehrlichiosis, *J. Infect. Dis.* **172:**1007–1012.
14. Whipple, G. H., 1907, A hitherto undescribed disease characterized anatomically by deposits of fat and fatty acids in the intestinal and mesenteric lymphatic tissues, *Johns Hopkins Hosp. Bull.* **18:**832–391.
15. Wilson, K. H., Blitchington, R., Frothingham, R., and Wilson, J. A., 1991, Phylogeny of the Whipple's-disease-associated bacterium, *Lancet* **338:**474–475.
16. Relman, D. A., Schmidt, T. M., MacDermott, R. P., and Falkow, S., 1992, Identification of the uncultured bacillus of Whipple's disease, *N. Engl. J. Med.* **327:**293–301.
17. Stoler, M. H., Bonfiglio, T. A., Steigbigel, R. T., and Pereira, M., 1983, An atypical subcutaneous infection associated with acquired immune deficiency syndrome, *Am. J. Clin. Pathol.* **80:**714–718.
18. Relman, D. A., Loutit, J. S., Schmidt, T. M., Falkow, S., and Tompkins, L. S., 1990, The agent of bacillary angiomatsosi: An approach to the identification of uncultured pathogens, *N. Engl. J. Med.* **23:**1573–1580.
19. Regnery, R. L., Anderson, B. E., Clarridge, J. E. III, Rodriquez-Barradas, M. C., Jones, D. C., and Carr, J. H., 1992, Characterization of a novel *Rochalimaea* species, *R. henselae* sp. nov., isolated from blood of a febrile, human immunodeficiency virus-positive patient, *J. Clin. Microbiol.* **30:**265–274.
20. Welch, D. F., Pickett, D. A., Slater, L. N., Steigerwalt, A. G., and Brenner, D. J., 1992, *Rochalimaea henselae* sp. nov., a cause of septicemia, bacillary angiomatosis, and parenchymal bacillary peliosis, *J. Clin. Microbiol.* **30:**275–280.
21. Brenner, D. J., O'Connor, S. P., Winkler, H. H., and Steigerwalt, A. G., 1993, Proposals to unify the genera *Bartonella* and *Rochalimaea,* with descriptions of *Bartonella quintana* comb. nov. *Bartonella vinsonii* comb. nov., *Bartonella henselae* comb. nov., and *Bartonella elizabethae* comb. nov., and to remove the family *Bartonellaceae* from the order *Rickettsiales, Int. J. Syst. Bacteriol.* **43:**777–786.
22. Anderson, B. E. and Neuman, M. A., 1996, The genus *Bartonella* as emerging human pathogens, *Clin. Microbiol. Rev.,* in press.
23. Anderson. B., Kelly, C., Threlkel, T., and Edwards, K., 1993, Detection of *Rochalimaea henselae* in cat-scratch disease skin test antigens, *J. Infect. Dis.* **168:**1034–1035.
24. Debre, R., Lamy, M., Jammet, M.-L., Costil, J., and Mozziconacci, P., 1950, La maladie des griffes de chat, *Bull. Mem. Soc. Med. Hop. Paris* **66:**76–79.
25. Suenaga, K., Okazaki, K., Yokoyama, Y., and Yamamoto, Y., 1995, Mycobacteria in the intestine of Japanese patients with inflammatory bowel disease, *Am. J. Gastroenterol.* **90:**76–80.
26. Shigematsu, N., Kido, K., Ikeda, T., Kamikawaji, N., and Sasazuki, T., 1994, Research trend for pathogenesis of sarcoidosis—streptococcal cell wall component, *Jpn. J. Clin. Med.* **52:**1498–1502.

27. Nakata, Y., Kataoka, M., and I Kimura., 1994, Sarcoidosis and *Propionibacterium acnes*. *Jpn. J. Clin. Med.* **52:**1492–1497.
28. Mangiapan, G. and Hance, A., 1995, Mycobacteria and sarcoidosis: An overview and summary of recent molecular biology data, *Sarcoidosis* **12:**20–37.
29. Tamura, N., Suzuki, K., Iwase, A., Yamamoto, T., and Kira, S., 1994, Retroviral infection as a putative pathogen for sarcoidosis, *Jpn. J. Clin. Med.* **52:**1492–1497.
30. Raoult, D., Fournier, P., Drancourt, M., Marrie, T., Etienne, J., Cosserat, J., Cacoub, P., Poinsignon, Y., Leclercq, P., and Sfeton, A., 1996, Diagnosis of 22 new cases of *Bartonella* endocarditis, Ann. Intern. Med. **125:**646–652.
31. Genetics Computer Group, 1994, Program manual for the Wisconsin package, version 8.0, 575 Science Drive, Madison WI 53711.

9

The Molecular Epidemiology of Nosocomial Infection

An Overview of Principles, Application, and Interpretation

RICHARD V. GOERING

1. INTRODUCTION

The treatment of infectious diseases centers around two general goals: (1) curing the patient and (2) preventing or at least restricting the spread of the disease. In a perfect world, health-care professionals would know that these goals have been achieved when (1) the patient is restored to good health and (2) there are no occurrences of newly infected patients. Unfortunately, the "real world" of infectious disease is far from perfect. Sooner or later the individual patient may present with evidence of additional infection by a given pathogen, often at a different body site. In a group of patients, more than one individual may yield cultures of the same species of organism. In both instances, the question commonly asked is whether multiple isolates of a given pathogen represent the same strain of organism. In the individual patient, this question commonly relates to issues of therapeutic efficacy whereas in a group of patients the concern is often one of infection

RICHARD V. GOERING • Department of Medical Microbiology and Immunology, Creighton University School of Medicine, Omaha, Nebraska 68178.

Rapid Detection of Infectious Agents, edited by Specter *et al.* Plenum Press, New York, 1998.

control. However, in both settings, the resolution of these questions is aided by specific epidemiological assessment.

2. THE BIRTH OF "MOLECULAR EPIDEMIOLOGY"

A variety of methods have been traditionally used in the epidemiological evaluation of nosocomial infections.[1–3] In general, the oldest approaches relied on a comparative assessment of phenotypic characteristics, such as biotype, serotype, and susceptibility to antimicrobial agents, various bacteriophages, or bacteriocins. Over time, efforts to obtain a more fundamental assessment of strain interrelationships have led to more molecularly based methods including comparisons of protein molecular weight distributions by polyacrylamide gel electrophoresis, relative mobility of specific enzymes by starch-gel electrophoresis (multilocus enzyme electrophoresis), specific antibody reactions with immobilized cellular proteins (immunoblotting), and cellular plasmid content (i.e., plasmid fingerprinting). These more molecular methods have clearly represented a technological step forward in epidemiologically assessing of infectious disease.

3. CHROMOSOMAL DNA: THE ULTIMATE MOLECULE OF EPIDEMIOLOGICAL COMPARISON

Chromosomal DNA represents the most fundamental molecule of identity in the cell. Thus, a variety of approaches to "molecular epidemiology" have been directed toward attaining at least some measure of the chromosomal relatedness of isolates.[1–3] These methods may be broadly categorized as (1) utilizing frequently cutting restriction enzymes to generate chromosomal restriction-fragment patterns for analysis [restriction-endonuclease analysis (REA)] by agarose gel electrophoresis, in some instances including hybridization with specific probes; (2) amplification of chromosomal sequences by PCR; and (3) analysis of chromosomal macro-restriction fragment patterns by pulsed field gel electrophoresis (PFGE). As illustrated in Fig. 1, frequently cutting restriction enzymes digest a megabase-sized DNA molecule, such as the bacterial chromosome, into a myriad of restriction-fragment sizes. When separated by agarose gel electrophoresis, complex banding patterns result [Fig. 1(a)] which are difficult to interpret comparatively. As shown in Fig. 1(b), the situation is considerably improved with the hybridization of probes to specific chromosomal sequences which, when detected, identify specific chromosomal restriction-fragment subsets. These subsets represent only a fraction of the total chromosomal DNA. Therefore, the discriminatory power of this approach is directly related to the degree of variability in chromo-

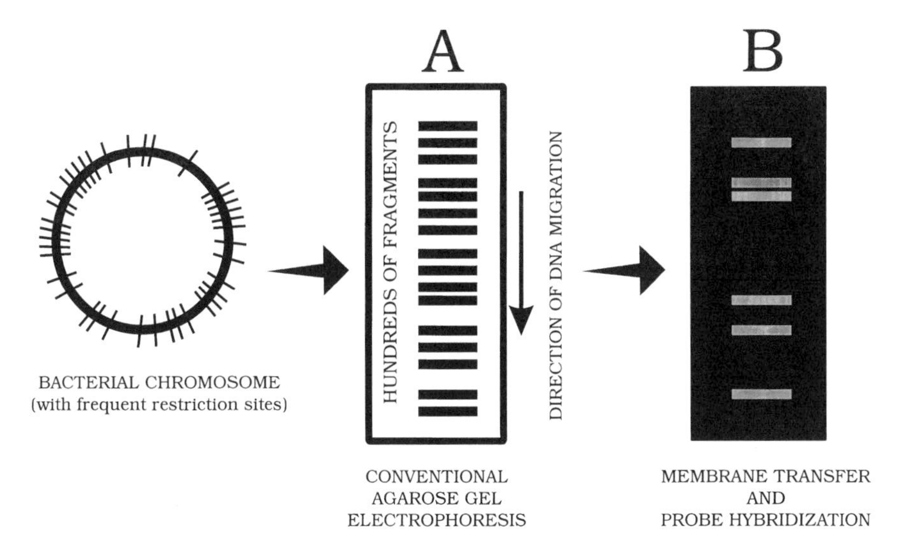

FIGURE 1. Diagrammatic representation of restriction-endonuclease analysis of chromosomal DNA (A) using conventional electrophoresis alone or (B) with subsequent hybridization using a specific probe.

somal regions flanking the probe-specific sequences. It is this potential variability that influences the size of the restriction fragments detected, commonly referred to as restriction fragment length polymorphism (RFLP) analysis. However, some chromosomal probes are not sufficiently discriminating for epidemiological purposes (e.g., highly conserved rRNA genes[4–7]). Conversely, other sequences have found epidemiological utility, as in the use of IS*6110* probes for monitoring *Mycobacterium tuberculosis*.[8]

A variety of amplification-based methods have been applied to the epidemiological analysis of nosocomial pathogens.[9] In one approach, a single primer of arbitrarily chosen DNA sequence is utilized to amplify random genomic regions [i.e., random amplified polymorphic DNA (RAPD) or arbitrarily primed PCR (AP-PCR)]. Alternatively, regions between repetitive chromosomal sequences (e.g., REP or BOX elements[10,11]) are amplified (termed rep-PCR), as shown in Fig. 2(a) and (b), respectively. In either case the goal is to generate amplicon sets which (when analyzed by agarose gel electrophoresis) vary between unrelated organisms and serve as genomic fingerprints. These methods have received widespread attention and increasing use in recent years principally because of their speed and apparent simplicity. However, several recent studies[9,12,13] have documented the many caveats associated with using amplification protocols, especially at low annealing temperatures, including the variable influence of virtually

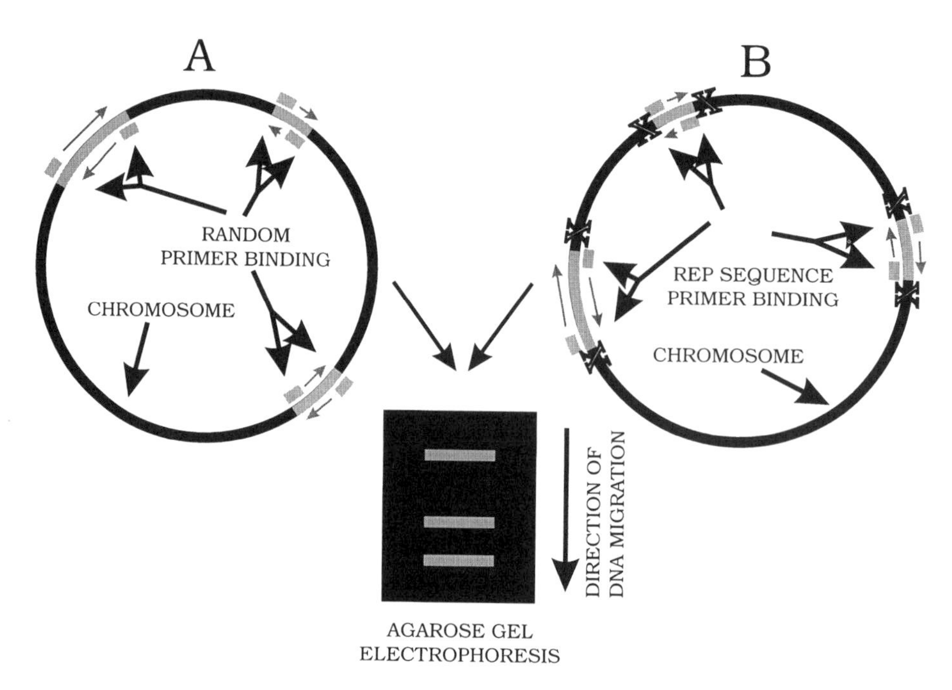

FIGURE 2. Diagrammatic representation of amplification methods where primers bind to allow amplification (A) of random genomic regions or (B) of regions between closely-spaced repetitive chromosomal sequences (indicated by an "X"). In either case, the results are analyzed by conventional agarose gel electrophoresis.

every component of the reaction mixture (e.g., polymerase, template, primer, and magnesium concentration), differences in thermocyclers, and difficulties with the databasing and interlaboratory comparison of information. In addition, as discussed above with probes, the amplicon sets generated by these methods survey and allow comparing only a small fraction of the total chromosomal DNA (Fig. 2).

In contrast to REA [Fig. 1(a)], exposure of the bacterial chromosome to a rare-cutting restriction enzyme results in a small number of DNA fragments (e.g., 7–20) the majority of which are greater than 50 kb in size. These macrorestriction fragments are too large to be separated by conventional agarose gel electrophoresis. But they can be resolved by PFGE[14] via a size-dependent response to the electrophoretic current which is "pulsed" in different directions for different lengths of time.[15] As shown in Fig. 3, the epidemiological "power" of PFGE lies in its unique ability to provide the investigator with a sense of the spatial distribution of a rare, repeated, specific (restriction) sequence around the entire bacterial chromosome. In addition, the method provides a unique degree of chromosomal

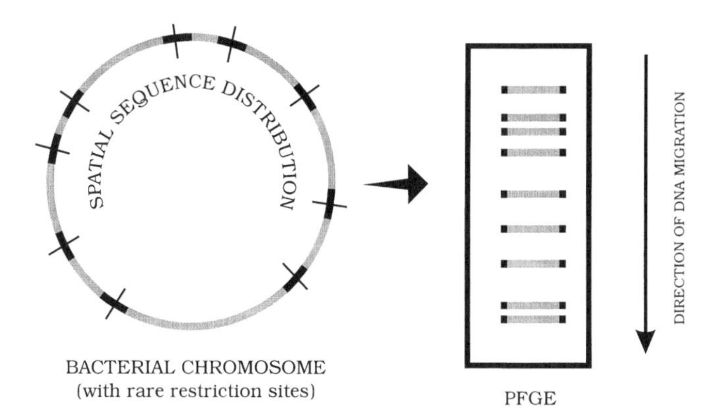

FIGURE 3. Diagrammatic representation of pulsed-field gel electrophoresis. A rare-cutting enzyme cleaves the bacterial chromosome into a small number of restriction fragments. The pattern of fragment sizes produced corresponds to the spatial distribution of the rare restriction sequence around the chromosome.

monitoring because a variety of genetic events may be detected as a change in the intrachromosomal distance between restriction-sequence copies (i.e., a change in restriction-fragment size). The importance of chromosomal monitoring for epidemiological subtyping is discussed in greater detail later. The current scientific literature demonstrates that PFGE is the method of choice (i.e., gold standard) for epidemiologically typing most prokaryotic and some eukaryotic etiological agents of infectious disease. At present, two exceptions are *M. tuberculosis*, where IS*6110* RFLP analysis is preferred[8] (see previous), and *Clostridium difficile* where nuclease activity in some isolates precludes definitive PFGE analysis[16,17] (then REA and AP-PCR are often utilized[18–20]). Nevertheless, to date, PFGE has been applied to the genetic and/or epidemiological analysis of at least 98 specific pathogens or pathogen groups (prokaryotic and eukaryotic) as shown in Table I.

4. MOLECULAR EPIDEMIOLOGICAL APPLICATIONS

As mentioned previously, molecular epidemiology is most commonly employed in two general scenarios: where multiple isolates of a given organism are cultured either from an individual patient or from a group of patients. In such instances, a variety of molecular approaches to epidemiological analysis may be utilized. However the data produced by PFGE are generally the most intuitive and "straightforward" to interpret.[157] Therefore, the discussion to follow focuses on PFGE as a paradigm of molecular epidemiology and its usefulness for strain typing in specific infection settings.

TABLE I
Organisms Analyzed by PFGE

Acinetobacter baumanii,[21,22] *A. calcoaceticus*[23]
Aeromonas hydrophila[24]
Alcaligenes xylosoxidans[25]
Bacteroides spp.[26]
Bartonella (Rochalimaea) spp.[27]
Bordetella parapertussis,[28] *B. pertussis*[29,30]
Borrelia burgdorferi[31,32]
Burkholderia (Pseudomonas) cepacia[25,33]
Campylobacter coli,[34,35] *C. fetus*,[36] *C. jejuni*,[34,35] *C. upsaliensis*[37]
Candida (Torulopsis) glabrata,[38,39] *C. albicans*,[40,41] *C. lusitaniae*,[42,43] *C. molischiana*,[44] *C. parapsilosis*,[45,46] *C. rugosa*,[47] *C. tropicalis*[48]
Chlamydia trachomatis[49,50]
Citrobacter koseri[51]
Clostridium acetobutylicum,[52] *C. botulinum*,[53] *C. difficile*,[18,19] *C. perfringens*[17,54]
Corynebacterium diptheriae[55]
Coxiella burnetti[56]
Cryptococcus neoformans[57,58]
Eimeria tenella[59]
Entamoeba histolytica[60]
Enterobacter aerogenes,[61] *E. cloacae*[62,63]
Enterococcus faecalis,[64,65] *E. faecium*[66,67]
Escherichia coli[68,69]
Flavobacterium meningosepticum[70]
Giardia duodenalis[71]
Haemophilus influenzae,[72,73] *H. parainfluenzae*[74]
Helicobacter mustelae,[75] *H. pylori*[76,77]
Klebsiella pneumoniae[78,79]
Legionella bozemanii,[80] *L. pneumophilia*[81,82]
Leptospira borgpetersenii,[80] *Leptospira spp.*[84]
Listeria monocytogenes[5,85]
Moraxella (Branhamella) catarrhalis[86,87]
Mycobacterium abscessus,[88,89] *M. avium*,[90,91] *M. bovis*,[92,93] *M. chelonae*,[89] *M. haemophilum*,[94] *M. kansasii*,[95] *M. paratuberculosis*,[96] *M. tuberculosis*[92,93]
Mycoplasma pneumoniae,[97] *M. hominis*[98]
Neisseria gonorrhoeae,[99,100] *N. meningitidis*[101,102]
Ochrobactrum anthropi[103]
Pasteurella multocida[104]
Plasmodium falciparum[105]
Pneumocystis carinii[106]
Pseudomonas aeruginosa[107,108]
Rickettsia akari,[109] *R. prowazekii*,[110] *R. typhi*[110]
Saccharomyces cerevisiae,[111,112] *S. spp.*[113]
Salmonella agona,[114,115] *S. brandenburg*,[116] *S. enterica*,[117,118] *S. enteritidis*,[119,120] *S. typhi*,[121,122] *S. typhimurium*,[123] *S. spp.*[124]
Serratia marcescens,[125,126] *S. odorifera*[127]

TABLE I *(Continued)*

Shigella dysenteriae,[128] *S. flexneri,*[129] *S. sonnei*[130,131]
Sporothrix schenckii[132]
Staphylococcus aureus,[133,134] *S. epidermidis,*[135,136] *S. haemolyticus,*[137] *S. lugdunensis*[138]
Stenotrophomonas (Xanthomonas) maltophilia[25,139]
Streptococci (group A),[140,141] *Streptococci (group B),*[142,143] *S. pneumoniae*[144,145]
Trypanosoma cruzi,[146] *T. rangeli*[147]
Vibrio anguillarum,[148] *V. cholerae,*[149,150] *V. parahaemolyticus,*[151] *V. vulnificus*[152,153]
Yersinia enterocolitica,[154,155] *Y. pestis*[156]

4.1. Infection in the Individual Patient

During antibiotic chemotherapy the individual patient may yield multiple cultures of a given pathogen with different antimicrobial susceptibilities. This commonly takes the form of a susceptible pretreatment isolate coupled with one or more posttreatment isolates that exhibit decreased susceptibility or antibiotic resistance. Two very different processes could produce this result: (1) emergence of resistance in the originally susceptible isolate (an issue of therapeutic efficacy) or (2) superinfection of the patient with a resistant organism unrelated to the pretreatment isolate (a matter of infection control). The usefulness of molecular epidemiology in clarifying such a situation is illustrated in Fig. 4. Here, PFGE analysis demonstrates that no relationship exists between a susceptible pretreatment and methicillin-resistant posttreatment isolate of *Staphylococcus aureus* [Fig. 4a, lanes 1 and 2, respectively]. Conversely, expanded spectrum beta-lactamase activity allowed an apparently "susceptible" *Enterobacter aerogenes* strain to emerge resistant to cephalosporin therapy [Fig. 4b, lanes 1 and 2, respectively].

In some instance, a patient receiving antimicrobial chemotherapy yields multiple isolates of a given organism with similar or identical antimicrobial susceptibility patterns. In such a setting, epidemiological analysis by a molecular method such as PFGE may provide a definitive statement on isolate interrelationships. This information is especially vital for blood isolates, where sequential cultures of the "same" strain of organism point toward bacteremia/septicemia, whereas sequential unrelated isolates would more be likely to represent contamination. In addition, an examination of interrelationships between phenotypically similar isolates cultured from different body sites can provide insight into whether a patient is infected with more than one strain of a given pathogen. Figure 4(c) illustrates such an analysis with three isolates of *S. aureus* that exhibit similar antibiograms but are cultured from different body sites in the same patient. When examined by PFGE, two sequential blood isolates (lanes 1 and 3) are clearly unrelated. However, a surgical wound isolate appears identical to the second blood isolate (lanes 2 and 3, respectively). Analysis of additional blood cultures

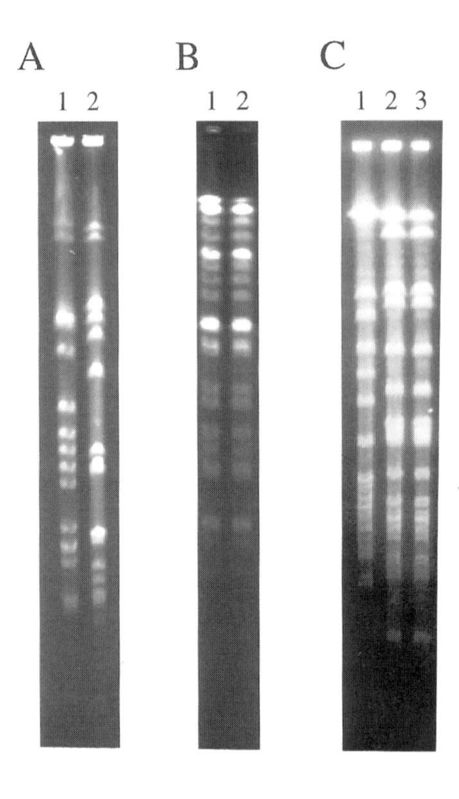

FIGURE 4. PFGE analysis of (A) *Sma*I-digested chromosomal DNA from methicillin-susceptible (lane 1) and methicillin-resistant (lane 2) *Staphylococcus aureus* isolates from an individual patient; (B) *Xba*I-digested chromosomal DNA from antibiotic-susceptible pretreatment (lane 1) and resistant postreatment (lane 2) *Enterobacter aerogenes* isolates from an individual patient; and (C) *Sma*-I digested chromosomal DNA of *S. aureus* isolated from the blood (lanes 1 and 3) and a surgical wound (lane 2) of an individual patient.

from the patient would aid in determining whether the two blood isolates are truly coinfecting or if one [e.g., Fig. 4(c), lane 1] is a contaminant.

4.2. Infection in Multiple Patients

There is always concern when more than one patient exhibits infection due to the same etiological agent, especially in a nosocomial setting. In this context, a number of scenarios suggest problems of infection control that could benefit from analysis by molecular typing. Specific "red flags" commonly include increased detection (i.e., frequency of isolation) of a particular pathogen or an increase in infections associated with a specific patient group. In addition, even in small numbers, some especially troublesome pathogens [e.g., *Pseudomonas* spp. in a neonatal intensive care unit or vancomycin-resistant enterococci (VRE)] are cause for concern. In such instances, epidemiological analysis including the use of a molecular method, such as PFGE, may reveal the extent to which patient-to-patient spread is a contributing factor. This is illustrated in Fig. 5, where VRE isolates from 12 different patients clearly represent a single "epidemic" strain.

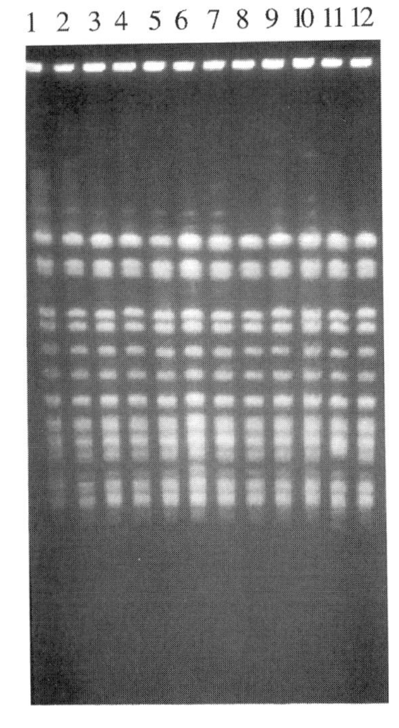

FIGURE 5. PFGE analysis of *Sma*I-digested chromosomal DNA from vancomycin-resistant *Enterococcus faecium* isolated from 12 different patients (lanes 1–12).

Such information allow more efficient, specific targeting of infection control measures to restrict further spread of the organism.

5. THE INTERPRETATION OF MOLECULAR TYPING DATA

5.1. Principles

In assessing the utility of various molecular typing methods, one of the most fundamental considerations is ease and reliability of interpretation. As discussed above, the newest typing methods aspire to achieve epidemiological comparisons based on some assessment of chromosomal relatedness. Thus, the central issue for interpretation becomes one of "significant degree of chromosomal difference." This can be considered in terms of three fundamental questions:

1. Is it possible that a change could occur within the chromosome of a pathogen as it transfers from one patient to another or from one body site to another in the same patient?
2. If the answer to question (1) is yes, what type of change(s) might occur?
3. How sensitive is a particular typing method in detecting such a change?

Regarding the first question, as an organism transfers from one site to another, one would certainly expect the majority of isolates to be essentially identical chromosomally and thus be recognized as the same "strain." However, although the chromosomes of microbial pathogens are certainly stable, they are by no means "cast in stone." Mutations may occur over time. Thus, as suggested by Tenover *et al.*,[157] it is reasonable to consider that an organism could sustain a single chromosomal change (i.e., a single genetic event) as it transfers to a different site. Though not identical, these epidemiologically related isolates would be viewed as "strain subtypes."

The response to the previous (2) question may be summarized in four categories of possible genetic change (to be discussed in greater detail later): insertions, deletions, rearrangements, and single-base substitutions (i.e., a transition or transversion). In considering these changes, attention must be paid to their potential effect on restriction sites because comparing restriction-fragment sizes is basic to several epidemiological approaches.

Question (3) raises a central issue regarding the efficacy of molecular typing which is considered in greater detail later. At present, PFGE is the most straightforward to interpret[3,158] of the chromosomally based typing methods currently in use. As noted previously, this results in major part from the ability of PFGE to provide a sense of global chromosomal monitoring via its potential to detect genetic events around the chromosome as they influence macrorestriction fragment sizes. For this reason, the following discussion centers on interpreting data generated by PFGE. However, the overall principles involved are generally applicable to any typing method which is based on comparing (i.e., size and number) chromosomal regions, such as restriction fragments or PCR-generated amplicons.

5.2. Chromosomal Changes as an Indicator of Epidemiological Relatedness

Figure 6 depicts the genetic changes listed previously, with specific reference to PFGE. Insertions, deletions, and rearrangements are depicted as they might, or might not, affect rare chromosomal restriction sites (Fig. 6, A1 and B1, A2 and B2, A3 and B3, respectively). Single-base substitutions (i.e., a transition or transversion) are not diagrammed but are discussed later. Figure 7 illustrates the influence of these genetic events on PFGE restriction-fragment patterns, depending on whether a restriction site is affected (Fig. 7A) or is not affected (Fig. 7B). The beginning "frame of reference" is the "epidemic" PFGE pattern in the middle of Fig. 7 which, in an actual nosocomial setting, would presumably have been observed in several isolates. In addition, one must remember the underlying principle of electrophoresis: the larger the DNA molecule, the slower it migrates in the electrophoretic field.

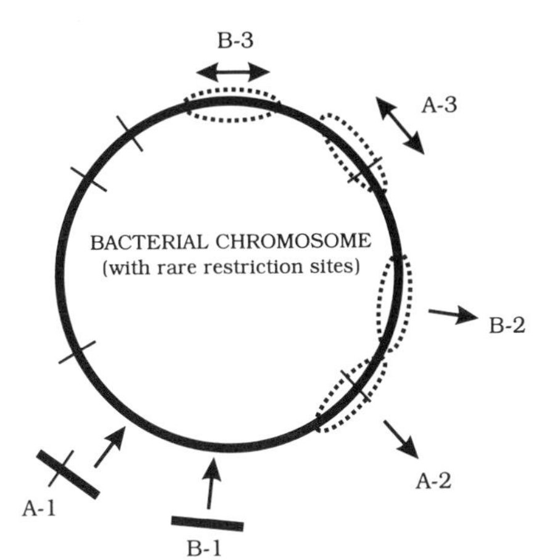

FIGURE 6. Diagrammatic representation of different chromosomal genetic events and their potential effect on PFGE interpretation. Insertions, deletions, and rearrangements are shown as they might or might not affect a rare chromosomal restriction site (A1 and B1, A2 and B2, A3 and B3, respectively).

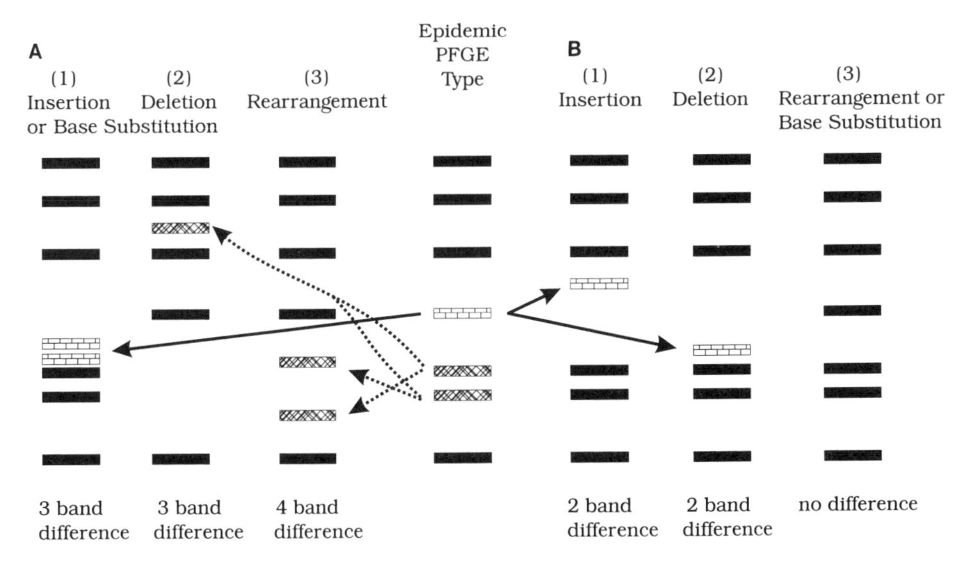

FIGURE 7. Diagrammatic representation of the influence of genetic events, shown in Fig. 6, on PFGE restriction-fragment patterns. Compared to the epidermic reference pattern, in the center, events are categorized as to whether (A) they would affect or (b) not affect a rare chromosomal restriction site.

5.2.1. Events Not Affecting a Chromosomal Restriction Site

Beginning first with genetic events which do not affect a restriction site, an insertion or deletion (Fig. 6, B1 and B2, respectively) would cause either a decrease or increase in the rate of electrophoretic migration of the affected restriction fragment (Fig. 7, B1 and B2, respectively). In either case, the result would appear as a difference in two restriction-fragment positions when compared to the "epidemic" PFGE pattern. A rearrangement (Fig. 6, B3) or base substitution not affecting a restriction site would have no influence on the PFGE pattern (Fig. 7, B3).

5.2.2. Events Affecting a Chromosomal Restriction Site

In genetic events affecting a restriction site, an insertion introducing a new restriction site would convert one restriction fragment into two (Fig. 6, A1). The result would be a difference in three restriction-fragment positions compared to the reference PFGE pattern (Fig. 7, A1). A deletion removing a restriction site would meld two adjacent restriction fragments into one (Fig. 6, A2), also producing a difference of three restriction-fragment positions (Fig. 7, A2). A single base substitution either creating or eliminating a restriction site would produce the same result as the insertions and deletions just discussed (Fig. 7, A1 and A2, respectively). A rearrangement relocating a restriction site would result in an increase in the size of one restriction fragment and a simultaneous decrease in the size of an adjacent restriction fragment (Fig. 6, A3) producing a difference of four restriction-fragment positions (Fig. 7, A3) compared to the "epidemic" PFGE type.

The reader will note that a genetic event leading to one difference in restriction-fragment position is not depicted in either Figs. 6 or 7 despite the fact that such observations are frequently reported in the scientific literature. The reason for this "omission" is that genetic events leading to a single restriction-fragment difference, while theoretically possible, would be extremely rare. For example, an event analogous to the precise insertion of a DNA fragment into a single chromosomal restriction site (in a manner preserving a functional copy of the restriction site at both "ends") would be required to produce one additional restriction fragment. Conversely, a deletion event, such as the excision of a restriction fragment precisely within its flanking restriction sequences (in a manner bringing together the chromosomal "ends" to produce a single functional restriction sequence) would result in the loss of a single restriction fragment. The relative rarity of such events suggests that most reported single restriction-fragment differences are actually a two-position difference where one of the fragments is simply hidden (i.e., comigrating with another restriction fragment) in one of the isolates.

5.2.3. The Three-Restriction-Fragment Difference as a Guideline

As illustrated previously, a single genetic event could produce from zero to four differences in restriction-fragment position compared with the "parental" PFGE type. The range of potential differences in restriction-fragment position due to more than one genetic event would obviously be multiples of these amounts. Therefore, employing the general categories of epidemiological relatedness suggested by Tenover *et al.*,[157] nosocomial isolates with up to four differences in restriction-fragment position (i.e., potentially caused by a single genetic event) would be potential strain subtypes. However, in considering degrees of epidemiological relatedness, one must remember that certain single genetic events (e.g., insertions or deletions not affecting restriction sites; Figs. 6 and 7, B1 and B2, respectively) result in two differences in restriction-fragment position. The cumulative effect of two such events (indicative of more marginal epidemiological relatedness[157]) could be a difference of four restriction-fragment positions, indistinguishable from a single rearrangement event [Figs. 6, A3 and 7, A3). The simplest of genetic events (i.e., a single base change) may generate a difference of three restriction-fragment positions (Fig. 7, A1 and A2). Therefore, when utilizing a typing method such as PFGE in a nosocomial setting, a conservative guideline is to consider that isolates differing by ≤ 3 restriction-fragment positions could have occurred via a single genetic event. Thus such isolates may represent epidemiologically related subtypes of the same strain. Conversely, isolates differing in the position of >3 restriction fragments could potentially represent a much more tenuous epidemiological interrelationship. This is illustrated in Fig. 8 with hypothetical PFGE data for 10 nosocomial isolates. In this example, the isolates in lanes 2, 5, and 9 are identical, thus representing the same strain, here arbitrarily termed type A1. In Fig. 8, lanes 1, 4, and 7 are significantly different from A1 (i.e., >3 restriction fragment positions) but are closely related to each other, most probably representing a second epidemiological group with two subtypes (B1 and B2). Similarly, lanes 3, 6, and 10 comprise a third group with two subtypes (C1 and C2), whereas lane 8 is unrelated to groups A, B, and C, and thus represents an "outlying" organism, termed type D1. It must be noted, however, that the previous discussion and guidelines[157] are intended to be applied to isolates in a specific environmental (i.e., nosocomial) setting where differences and similarities pertain directly to epidemiological interrelationships. In a broader environment (e.g., countrywide or global surveys), one moves from epidemiology to more of a population biological context. In this latter case, similarly-sized restriction fragments in isolates of disparate origin may not necessarily represent equivalent chromosomal regions.[159] In such a setting, interpretive criteria must be applied with caution, as in the case of *Escherichia coli* 0157:H7 analysis where in some instances even an apparent single restriction-fragment difference may be signifi-

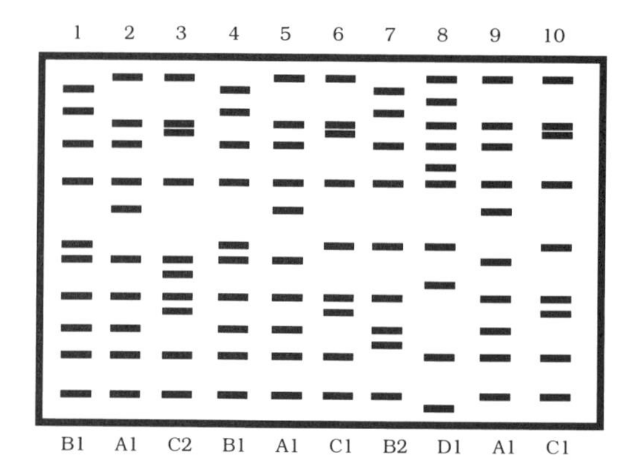

FIGURE 8. Diagrammatic representation of hypothetical PFGE data for 10 nosocomial isolates (lanes 1–10). Potential epidemiological interrelationships are illustrated utilizing a three-restriction-fragment difference as a guideline. PFGE patterns with the same alphabetical but different numerical designations are considered related subtypes.

cant.[160] In actual practice, however, it is important to stress that the epidemiological analysis of nosocomial infections should not rely on molecular data alone. The most accurate assessment of epidemiological interrelationships is achieved when molecular typing, by whatever method, is employed and interpreted in the context of all available clinical information (e.g., from the clinical microbiological laboratory, patient information, etc.).

5.2.4. *The Issue of Detectable Significant Change*

If one accepts degree of chromosomal change as an indicator of epidemiological relatedness, the ability to detect such genetic events becomes a central issue in considering the efficacy of different approaches to molecular typing. In this context, one might argue that the ultimate basis for epidemiological comparison is total chromosomal sequencing. However, random replication mistakes would most certainly lead to less than 100% base-for-base identity in a single organism's chromosome after a minimum number of generations. Thus, even if feasible, the concern with this approach would be one of "oversensitivity." Because less than 100% chromosomal identity is not automatically equivalent to epidemiological difference, the question becomes one of defining significant chromosomal change. In many respects, this represents a "gray area" for many nosocomial pathogens where fundamental genetic information (e.g., innate genomic stability and other parameters nicely summarized by Struelens *et al.*[3]) is lacking.

For epidemiological purposes, the issue of significant chromosomal change has two key aspects: (1) the frequency of the genetic event and (2) its ability to be detected. With regard to frequency, Tenover *et al.*[157] have proposed a good working hypothesis which, for a given organism, simply correlates increased numbers of chromosomal changes with decreased probability of epidemiological relatedness. Thus, the genetic events most suitable for epidemiological monitoring would be those whose frequency, on average, dictates only a limited number of cumulative chromosomal changes (i.e., ideally 0 to 1) as an organism circulates, at least for the short term, in a given (nosocomial) environment. This would logically include the genetic events discussed previously, which often result from chromosomal interaction with variety of extra chromosomal elements (e.g., transposons, plasmids, and bacteriophages[161,162]). In comparing chromosomal typing methods with regard to detection, it is obvious that genetic events will be recognized only within the chromosomal area being surveyed. Unfortunately, with most methods, it is presently unclear how much of the chromosome is actually being analyzed. With amplification-based typing methods, for example, the size of the amplicons generated represents only a fraction of the total chromosome, often with a relatively unknown spatial distribution. Thus, the discriminatory ability of these approaches lies in their ability to detect even minor changes in the small, but epidemiologically relevant, chromosomal regions they survey (e.g., interrepetitive sequences or intergenic spacer regions). In contrast, as noted before, PFGE is unique in imparting a sense of the spatial distribution of a rare, repeated (restriction) sequence round the entire bacterial chromosome. This is visualized as a pattern of macrorestriction fragments commonly ranging from >500 kb to <50 kb in size. Although the smaller restriction fragments (<50 kb) are generally more difficult to discern, the sum of the fragments >50 kb in size equates to a survey of >90% of the bacterial chromosome.[163–165] In general, genetic events (e.g, insertions and deletions) must affect at least 8% of restriction-fragment length to produce observable shifts in electrophoretic mobility with PFGE.[161,162] Thus the ability of PFGE to register "minor" chromosomal changes decreases with increasing restriction-fragment size. Nevertheless, even in the largest restriction fragments, major genetic changes, such as prophage insertion or deletion,[161,162] are detected. Thus, the method provides a unique sense of global chromosomal monitoring in its potential to detect genetic events around the chromosomes as they influence macrorestriction fragment sizes.

6. SUMMARY AND CONCLUSIONS

In the final analysis, there is no perfect molecular typing method. Each approach has is advantages and disadvantages. For example, of all the molecular typing methods available, PFGE and the amplification-based methods are clearly the best documented and have been applied to the greatest variety of nosocomial pathogens. But there are clear differences between the two.

In general, the amplification-based methods require reasonably inexpensive equipment and are relatively simple to perform. However, the "danger" in this apparent simplicity (noted by Tyler *et al.*[9]) is that essentially every component of the preamplification mixture has the potential to directly influence the data generated (i.e., size and number of amplicons). Thus, a lack of careful attention to seemingly simple preamplification details may result in a different, and potentially incorrect, epidemiological answer. In addition, the potential variability of the raw data generated by amplification methods is problematic for databasing and interlaboratory comparison.[9] It must also be remembered that the amplification-based methods represent a group of protocols which differ in the extent to which their discriminatory power has been validated with different nosocomial pathogens. As mentioned previously, amplification-based typing concentrates on detecting genetic differences in relatively small (but epidemiologically relevant) chromosomal regions.

Conversely, PFGE requires relatively more expensive equipment and is more complex to perform. However, recent years have seen dramatic reductions in DNA preparation time[166] including commercial streamlining into a "kit" format. In addition, PFGE is relatively "forgiving" of minor changes in DNA preparation protocol. In the event of a "mistake," the result is most likely to be recognized as incompletely digested restriction fragments or no data at all. As discussed before, the method uniquely provides a sense of global chromosomal monitoring in detecting a variety of genetic events around >90% of the chromosome, as they influence macrorestriction fragment sizes. Thus, at present, the current scientific literature suggests that PFGE is the gold standard for epidemiologically typing most nosocomial pathogens.

A comparison of molecular typing methods underscores the point that, in many respects, molecular epidemiology involves a series of choices where one option is "traded" for another. Such choices often translate into a question of time versus information. In a purely academic setting, one may have the time and intellectual focus to pursue molecular answers to questions of epidemiological relatedness to the nth degree. This might include the use of many different typing methods and multiple options within each method (e.g., the use of many different restriction enzymes, primer sets, etc.). Conversely, just the opposite is true of the clinical setting, where multiple focus and time constraints are the rule of the day. As a standard epidemiological protocol, one could consider utilizing two molecular methods, for example, beginning first with PFGE followed by an amplification-based approach for a final analysis of "problem" isolates. However, time constraints often dictate the use of only a single molecular method which one would choose for its ability to provide the greatest amount of epidemiological discrimination in the fewest number of tries. In addition, the number of options one might exercise within a given method are also subject to time pressure. For example, a review of the literature suggests that using a second or third restriction enzyme with PFGE rarely, if ever, changes the epidemiological interpretation (as discussed before). Thus, one must weigh whether repeated molecular typing (e.g., by PFGE) of the same isolates is worth the cost in time and effort.

The years ahead may ultimately see the development of an absolute (e.g., molecular) indicator of strain relatedness. Presently, however, the epidemiological analysis of nosocomial infections must continue to rely, as it always has, on the "educated guess." It is clear that the fundamental information provided by molecular typing has, and will continue, to facilitate this process. Nevertheless, it is important to restate the caveat that the epidemiological analysis of nosocomial infections should not center on molecular typing alone. In the final analysis, the most accurate assessment of epidemiological interrelationships in a nosocomial setting is achieved when all available information is considered together. Molecular typing brings a powerful component to this endeavor, which is no mere academic exercise in its proven potential to significantly affect health care worldwide.

REFERENCES

1. Arbeit, R. D., 1997, Laboratory procedures for epidemiologic analysis, in: *The Staphylococci in Human Disease* (K. B. Crossley and G. L. Archer, eds.), Churchill Livingstone, New York, pp. 253–286.
2. Maslow, J. and Mulligan, M. E., 1996, Epidemiologic typing system, *Infect. Control. Hosp. Epidemiol.* **17:**595.
3. Struelens, M. J., Gerner-Smidt, P., Rosdahl, V., El Solh, N., Etienne, J., Nicolas, M. H., Römling, U., Witte, W., Legakis, N., Van Belkum, A., Dijkshoorn, L., De Lencastre, H., Garaizar, J., Blanc, D., Bauernfeind, A., Cookson, B. D., and Pitt, T. L., 1996, Consensus guidelines for appropriate use and evaluation of microbial epidemiological typing systems, *Clin. Microbiol. Infect* **2:**1.
4. Gordillo, M. E., Singh, K. V., and Murray, B. E., 1993, Comparison of ribotyping and pulsed-field gel electrophoresis for subspecies differentiation of strains of *Enterococcus faecalis, J. Clin. Microbio.,* **31:**1570.
5. Louie, M., Jayaratne, P., Luchsinger, I., Devenish, J., Yao, J., Schlech, W., and Simor, A., 1996, Comparison of ribotyping, arbitrarily primed PCR, and pulsed-field gel electrophoresis for molecular typing of *Listeria monocytogenes, J. Clin. Microbiol.* **34:**15.
6. Martin, I. E., Tyler, S. D., Tyler, K. D., Khakhria, R., and Johnson, W. M., 1996, Evaluation of ribotyping as epidemiologic tool for typing *Escherichia coli* serogroup O157 isolates, *J. Clin. Microbiol.* **34:**720.
7. Seifert, H. and Gerner-Smidt, P., 1995, Comparison of ribotyping and pulsed-field gel electrophoresis for molecular typing of *Acinetobacter* isolates, *J. Clin. Microbiol.* **33:**1402.
8. Van Embden, J. D. A., Cave, M. D., Crawford, J. T., Dale, J. W., Eisenach, K. D., Gicquel, B., Hermans, P., Martin, C., McAdam, R., Shinnick, T. M., and Small, P. M., 1993, Strain identification of *Mycobacterium tuberculosis* by DNA fingerprinting: Recommendations for a standardized methodology, *J. Clin. Microbiol.* **31:**406.
9. Tyler, K. D., Wang, G., Tyler, S. D., and Johnson, W. M., 1997, Factors affecting reliability and reproducibility of amplification-based DNA fingerprinting of representative bacterial pathogens, *J. Clin. Microbiol.* **35:**339.
10. Versalovic, J., Koeuth, T., and Lupski, J. R., 1991, Distribution of repetitive DNA sequences in eubacteria and application to fingerprinting of bacterial genomes, *Nucleic Acid Res.* **19:**6823.

11. Martin, B., Humbert, O., Camara, M., Guenzi, E., Walker, J., Mitchell, T., Andrew, P., Prudhomme, M., Alloing, G., Hakanbeck, R., Morrison, D. A., Boulnois, G. J., and Claverys, J., 1992, A highly conserved repeated DNA element located in the chromosome of *Streptococcus pneumoniae*, *Nucleic Acids Res.* **20:**3479.
12. Van Belkum, A., Sluijter, M., De Groot, R., Verbrugh, H., and Hermans, P. W. M., 1996, Novel BOX repeat PCR assay for high-resolution typing of *Streptococcus pneumoniae* strains, *J. Clin. Microbiol.* **34:**1176.
13. Halldén, C., Hansen, M., Nilsson, N. O., Hjerdin, A., and Säll, T., 1996, Competition as a source of errors in RAPD analysis, *Theor. Appl. Genet.* **93:**1185.
14. Schwartz, D. C., Saffran, W., Welsh, J., Haas, R., Goldenberg, M., and Cantor, C. R., 1983, New techniques for purifying large DNAs and studying their properties and packaging, *Cold Spring Harbor Symp. Quant. Biol.* **47:**189.
15. Goering, R. V., 1993, Molecular epidemiology of nosocomial infection: analysis of chromosomal restriction patterns by pulsed field gel electrophoresis, *Infect. Control. Hosp. Epidemiol.* **14:**595.
16. Kato, H., Kato, N., Watanabe, K., Ueno, K., Ushijima, H., Hashira, S., and Abe, T., 1994, Application of typing by pulsed-field gel electrophoresis to the study of *Clostridium difficile* in a neonatal intensive care unit, *J. Clin. Microbiol.* **32:**2067.
17. Kristjánsson, M., Samore, M. H., Gerding, D. N., DeGirolami, P. C., Bettin, K. M., Karchmer, A. W., and Arbeit, R. D., 1994, Comparison of restriction endonuclease analysis, ribotyping, and pulsed-field gel electrophoresis for molecular differentiation of *Clostridium difficile* strains, *J. Clin. Microbiol.* **32:**1963.
18. Samore, M. H., Kristjansson, M., Venkataraman, L., DeGirolami, P. C., and Arbeit, R. D., 1996, Comparison of arbitrarily primed polymerase chain reaction, restriction enzyme analysis and pulsed-field gel electrophoresis for typing *Clostridium difficile*, *J. Microbiol. Methods* **25:**215.
19. Van Dijck, P., Avesani, V., and Delmée, M., 1996, Genotyping of outbreak-related and sporadic isolates of *Clostridium difficile* belonging to serogroup C, *J. Clin. Microbiol.* **34:**3049.
20. Collier, M. C., Stock, F., DeGirolami, P. C., Samore, M. H., and Cartwright, C. P., 1996, Comparison of PCR-based approaches to molecular epidemiologic analysis of *Clostridium difficile*, *J. Clin. Microbiol.* **34:**1153.
21. Kaul, R., Burt, J. A., Cork, L., Dedier, H., Garcia, M., Kennedy, C., Brunton, J., Krajden, M., and Conly, J., 1996, Investigation of a multiyear multiple critical care unit outbreak due to relatively drug-sensitive *Acinetobacter baumannii:* Risk factors and attributable mortality, *J. Infect. Dis.* **174:**1279.
22. Scerpella, E. G., Wanger, A. R., Armitige, L., Anderlini, P., and Ericsson, C. D., 1995, Nosocomial outbreak caused by a multiresistant clone of *Acinetobacter baumannii:* Results of the case-control and molecular epidemiologic investigations, *Infect. Control. Hops. Epidemiol.* **16:**92.
23. Allardet-Servent, A., Bouziges, N., Carles-Nurit, M.-J., Bourg, G., Gouby, A., and Ramuz, M., 1989, Use of low-frequency-cleavage restriction endonucleases for DNA analysis in epidemiological investigations of nosocomial bacterial infections, *J. Clin. Microbiol.* **27:**2057.
24. Talon, D., Dupont, M. J., Lesne, J., Thouverez, M., and Michel-Briand, Y., 1996, Pulsed-field gel electrophosis as an epidemiological tool for clonal identification of *Aeromonas hydrophila*, *J. Appl. Bacteriol.* **80:**277.
25. Vu-Thien, H., Moissenet, D., Valcin, M., Dulot, C., Tournier, G., and Garbarg-Chenon, A., 1996, Molecular epidemiology of *Burkholderia cepacia*, *Stenotrophomonas maltophilia*, and *Alcaligenes xylosoxidans* in a cystic fibrosis center, *Eur. J. Clin. Microbiol. Infect. Dis.* **15:**876.
26. Bedzyk, L. A., Shoemaker, N. B., Young, K. E., and Salyers, A. A., 1992, Insertion and excision of *Bacteroides* conjugative chromosomal elements, *J. Bacteriol.* **174:**166.
27. Roux, V. and Raoult, D., 1995, Inter- and intraspecies identification of *Bartonella (Rochalimaea) species*, *J. Clin. Microbiol.* **33:**1573.
28. Porter, J. F., Connor, K., and Donachie, W., 1996, Differentiation between human and ovine

isolates of *Bordetella parapertussis* using pulsed-field gel electrophoresis, *FEMS Microbiol. Lett.* **135:**131.

29. Beall, B., Cassiday, P. K., and Sanden, G. N., 1995, Analysis of *Bordetella pertussis* isolates from an epidemic by pulsed-field gel electrophoresis, *J. Clin. Microbiol.* **33:**3083.
30. Moissenet, D., Valcin, M., Marchand, V., Grimprel, E., Bégué, P., Garbarg-Chenon, A., and Vu-Thien, H., 1996, Comparative DNA analysis of *Bordetella pertussis* clinical isolates by pulsed-field gel electrophoresis, randomly amplified polymorphism DNA, and ERIC polymerase chain reaction, *FEMS Microbiol. Lett.* **143:**127.
31. Anderson, J. F., Flavell, R. A., Magnarelli, L. A., Barthold, S. W., Kantor, F. S., Wallich, R., Persing, D. H., Mathiesen, D., and Fikrig, E., 1996, Novel *Borrelia burgdorferi* isolates from *Ixodes scapularis* and *Ixodes dentatus* ticks feeding on humans, *J. Clin. Microbiol.* **34:**524.
32. Busch, U., Hizo-Teufel, C., Boehmer, R., Fingerle, V., Nitschko, H., Wilske, B., and Preac-Mursic, V., 1996, Three species of *Borrelia burgdorferi* sensu lato (*B. burgdorferi* sensu stricto, *B. afzelii*, and *B. garinii*) identified from cerebrospinal fluid isolates by pulsed-field gel electrophoresis and PCR, *J. Clin. Microbiol.* **34:**1072.
33. Ryley, H. C., Ojeniyi, B., Hoiby, N., and Weeks, J., 1996, Lack of evidence of nosocomial cross-infection by *Burkholderia cepacia* among Danish cystic fibrosis patients, *Eur. J. Clin. Microbiol. Infect. Dis.* **15:**755.
34. Taylor, D. E., Eaton, M., Yan, W., and Chang, N., 1992, Genome maps of *Campylobacter jejuni* and *Camphylobacter coli*, *J. Bacteriol.* **174:**2332.
35. Yan, W., Chang, N., and Taylor, D. E., 1991, Pulsed-field gel electrophoresis of *Campylobacter jejuni* and *Campylobacter coli* genomic DNA and its epidemiologic application, *J. Infect. Dis.* **163:**1068.
36. Rennie, R. P., Strong, D., Taylor, D. E., Salama, S. M., Davidson, C., and Tabor, H., 1994, *Campylobacter fetus* diarrhea in a hutterite colony: Epidemiological observations and typing of the causative organism, *J. Clin. Microbiol.* **32:**721.
37. Bourke, B., Sherman, P. M., Woodward, D., Lior, H., and Chan, V. L., 1996, Pulsed-field gel electrophoresis indicates genotypic heterogeneity among *Campylobacter upsaliensis* strains, *FEMS Microbiol. Lett.* **143:**57.
38. Cormican, M. G., Hollis, R. J., and Pfaller, M. A., 1996, DNA macrorestriction profiles and antifungal susceptibility of *Candida (Torulopsis) glabrata*, *Diagn. Microbiol. Infect. Dis.* **25:**83.
39. Vazquez, J. A., Beckley, A., Donabedian, S., Sobel, J. D., and Zervos, M. J., 1993, Comparison of restriction enzyme analysis versus pulsed-field gradient gel electrophoresis as a typing system for *Torulopsis glabrata* and *Candida* species other than *C. albicans*, *J. Clin. Microbiol.* **31:**2021.
40. Barchiesi, F., Hollis, R. J., Del Poeta, M., McGough, D. A., Scalise, G., Rinaldi, M. G., and Pfaller, M. A., 1995, Transmission of fluconazole-resistant *Candida albicans* between patients with AIDS and oropharyngeal candidiasis documented by pulsed-field gel electrophoresis, *Clin. Infect. Dis.* **21:**561.
41. Porter, S. D., Noble, M. A., and Rennie, R., 1996, A single strain of *Candida albicans* associated with separate episodes of fungemia and meningitis, *J. Clin. Microbiol.* **34:**1813.
42. King, D., Rhine-Chalberg, J., Pfaller, M. A., Moser, S. A., and Merz, W. G., 1995, Comparison of four DNA-based methods for strain delineation of *Candida lusitaniae*, *J. Clin. Microbiol.* **33:**1467.
43. Pfaller, M. A., Messer, S. A., and Hollis, R. J., 1994, Strain delineation and antifungal susceptibilities of epidemiologically related and unrelated isolates of *Candida lusitaniae*, *Diagn. Microbiol. Infect. Dis.* **20:**127.
44. Janbon, G., Magnet, R., Bigey, F., Arnaud, A., and Galzy, P., 1995, Karyotype studies on different strains of *Candida molischiana* by pulsed-field gel electrophoresis, *Curr. Genet.* **28:**150.
45. Pfaller, M. A., Messer, S. A., and Hollis, R. J., 1995, Variations in DNA subtype, antifungal susceptibility, and slime production among clinical isolates of *Candida parapsilosis*, *Diagn. Microbiol. Infect. Dist.* **21:**9.

46. Pontieri, E., Gregori, L., Gennarelli, M., Ceddia, T., Novelli, G., Dallapiccola, B., De Bernardis, F., and Carruba, G., 1996, Correlation of *Sfi*I macrorestriction endonuclease fingerprint analysis of *Candida parapsilosis* isolates with source of isolation, *J. Med. Microbiol.* **45:**173.
47. Dib, J. C., Dube, M., Kelly, C., Rinaldi, M. G., and Patterson, J. E., 1996, Evaluation of pulsed-field gel electrophoresis as a typing system for *Candida rugosa:* Comparison of karotype and restriction fragment length polymorphisms, *J. Clin. Microbiol.* **34:**1494.
48. Doebbeling, B. N., Lehmann, P. F., Hollis, R. J., Wu, L.-C., Widmer, A. F., Voss, A., and Pfaller, M. A., 1993, Comparison of pulsed-field gel electrophoresis with isoenzyme profiles as a typing system for *Candida tropicalis, Clin. Infect. Dis.* **16:**377.
49. Birkelund, S. and Stephens, R. S., 1992, Construction of physical and genetic maps of *Chlamydia trachomatis* serovar L2 by pulsed-field gel electrophoresis, *J. Bacteriol.* **174:**2742.
50. Rodriguez, P., Allardet-Servent, A., De Barbeyrac, B., Ramuz, M, and Bebear, C., 1994, Genetic variability among *Chlamydia trachomatis* reference and clinical strains analyzed by pulsed-field gel electrophoresis, *J. Clin. Microbiol.* **32:**2921.
51. Papasian, C. J., Kinney, J., Coffman, S., Hollis, R. J., and Pfaller, M. A., 1996, Transmission of *Citrobacter koseri* from mother to infant documented by ribotyping and pulsed-field gel electrophoresis, *Diagn. Microbiol. Infect. Dis.* **26:**63.
52. Wilkinson, S. R. and Young, M., 1993, Wide diversity of genome size among different strains of *Clostridium acetobutylicum, J. Gen. Microbiol.* **139:**1069.
53. Lin, W. J. and Johnson, E. A., 1995, Genome analysis of Clostridium botulinum type A by pulsed-field gel electrophoresis, *Appl. Environ. Microbiol.* **61:**4441.
54. Katayama, S., Dupuy, B., Daube, G., China, B., and Cole, S. T., 1996, Genome mapping of *Clostridium perfringens* strains with I-*Ceu*I shows many virulence genes to be plasmid-borne, *Mol. Gen. Genet.* **251:**720.
55. De Zoysa, A., Efstratiou, A., George, R. C., Jahkola, M., Vuopio-Varkila, J., Deshevoi, S., Tseneva, G., and Rikushin, Y., 1995, Molecular epidemiology of *Corynebacterium diphtheriae* from northwestern Russia and surrounding countries studied by using ribotyping and pulsed-field gel electrophoresis, *J. Clin. Microbiol.* **33:**1080.
56. Heinzen, R., Stiegler, G. L., Whiting, L. L., Schmitt, S. A., Mallavia, L. P., and Frazier, M. E., 1990, Use of pulsed field gel electrophoresis to differentiate *Coxiella burnetii* strains, *Ann. N.Y. Acad. Sci.* **590:**504.
57. Dromer, F., Varma, A., Ronin, O., Mathoulin, S., and Dupont, B., 1994, Molecular typing of *Cryptococcus neoformans* serotype D clinical isolates, *J. Clin. Microbiol.* **32:**2364.
58. Fries, B. C., Chen, F. Y., Currie, B. P., and Casadevall, A., 1996, Karyotype instability in *Cryptococcus neoformans* infection, *J. Clin. Microbiol.* **34:**1531.
59. Shirley, M. W., Kemp, D. J., Pallister, J., and Prowse, S. J., 1990, A molecular karyotype of *Eimeria tenella* as revealed by contour-clamped homogeneous electric field gel electrophoresis, *Mol. Biochem. Parasitol.* **38:**169.
60. Petter, R., Rozenblatt, S., Schechtman, D., Wellems, T. E., and Mirelman, D., 1993, Electrophoretic karyotype and chromosome assignments for a pathogenic and a nonpathogenic strain of *Entamoeba histolytica, Infect. Immun.* **61:**3574.
61. De Gheldre, Y., Maes, N., Rost, F., De Ryck, R., Clevenbergh, P., Vincent, J. L., and Struelens, M. J., 1997, Molecular epidemiology of an outbreak of multidrug-resistant *Enterobacter aerogenes* infections and *in vivo* emergence of imipenem resistance, *J. Clin. Microbiol.* **35:**152.
62. Haertl, R. and Bandlow, G., 1993, Epidemiological fingerprinting of *Enterobacter cloacae* by small-fragment restriction endonuclease analysis and pulsed-field gel electrophoresis of genomic restriction fragments, *J. Clin. Microbiol.* **31:**128.
63. Shi, Z. Y., Liu, P. Y. F., Lau, Y. J., Lin, Y. H., and Hu, B. S., 1996, Epidemiological typing of isolates from an outbreak of infection with multidrug-resistant *Enterobacter cloacae* by repetitive

extragenic palindromic unit b1-primed PCR and pulsed-field gel electrophoresis, *J. Clin. Microbiol.* **34:**2784.

64. Seetulsingh, P. S., Tomayko, J. F., Coudron, P. E., Markowitz, S. M., Skinner, C., Singh, K. V., and Murray, B. E., 1996, Chromosomal DNA restriction endonuclease digestion patterns of β-lactamase-producing the *Enterococcus faecalis* isolates collected from a single hospital over a 7-year period, *J. Clin. Microbiol.* **34:**1892.
65. Tomayko, J. F. and Murray, B. F., 1995, Analysis of *Enterococcus faecalis* isolates from intercontinental sources by multilocus enzyme electrophoresis and pulsed-field gel electrophoresis, *J. Clin. Microbiol.* **33:**2903.
66. Montecalvo, M. A., Horowitz, H., Gedris, C., Carbonaro, C., Tenover, F. C., Issah, A., Cook, P., and Wormser, G. P., 1994, Outbreak of vancomycin-, ampicillin-, and aminoglycoside-resistant *Enterococcus faecium* bacteremia in an adult oncology unit, *Antimicrob. Agents Chemother.* **38:**1363.
67. Moreno, F., Grota, P., Crisp, C., Magnon, K., Melcher, G. P., Jorgensen, J. H., and Patterson, J. E., 1995, Clinical and molecular epidemiology of vanomycin-resistant *Enterococcus faecium* during its emergence in a city in southern Texas, *Clin. Infect. Dis.* **21:**1234.
68. Krause, U., Thomson-Carter, F. M., and Pennington, T. H., 1996, Molecular epidemiology of *Escherichia coli* O157:H7 by pulsed-field gel electrophoresis and comparison with that by bacteriophage typing, *J. Clin. Microbiol.* **34:**959.
69. Oethinger, M., Conrad, S., Kaifel, K., Cometta, A., Bille, J., Klotz, G., Glauser, M. P., Marre, R., and Kern, W. V., 1996, Molecular epidemiology of fluoroquinolone-resistant *Escherichia coli* bloodstream isolates from patients admitted to European cancer centers, *Antimicrob. Agents. Chemother.* **40:**387.
70. Sader, H. S., Jones, R. N., and Pfaller, M. A., 1995, Relapse of catheter-related *Flavobacterium meningosepticum* bacteremia demonstrated by DNA macrorestriction analysis, *Clin. Infect. Dis.* **21:**997.
71. Sarafis, K. and Isaac-Renton, J., 1993, Pulsed-field gel electrophoresis as a method of biotyping of *Giardia duodenalis, AM. J. Trop. Med. Hyg.* **48:**134.
72. Curran, R., Hardie, K. R., and Tower, K. J., 1994, Analysis by pulsed-field gel electrophoresis of insertion mutations in the transferrin-binding system of *Haemophilus influenzae* type b, *J. Med. Microbiol.* **41:**120.
73. Musser, J. M., Kroll, J. S., Granoff, D. M., Moxon, E. R., Brodeur, B. R., Campos, J., Dabernat, H., Frederiksen, W., Hamel, J., Hammond, G., Hoiby, E. A., Jonsdottir, K. E., Kabeer, M., Kallings, I., Khan, W. N., Kilian, M., Knowles, K., Koornhof, H. J., Law, B., Li, K. I., Montgomery, J., Pattison, P. E., Piffaretti, J.-C., and Takala, A. K., 1990, Global genetic structure and molecular epidemiology of encapsulated *Haemophilus influenzae, Rev. Infect. Dis.* **12:**75.
74. Kauc, L. and Goodgal, S. H., 1989, The size and a physical map of the chromosome of *Haemophilus parainfluenzae, Gene* **83:**377.
75. Taylor, D. E., Chang, N., Taylor, N. S., and Fox, J. G., 1994, Genome conservation in *Helicobacter mustelae* as determined by pulsed-field gel electrophoresis, *FEMS Microbiol. Lett.* **118:**31.
76. Hirschl, A. M., Richter, M., Makristathis, A., Prückl, P. M., Willinger, B., Schütze, K., and Rotter, M. L., 1994, Single and multiple strain colonization in patients with *Helicobacter pylori*-associated gastritis: Detection by macrorestriction DNA analysis, *J. Infect. Dis.* **170:**473.
77. Salama, S. M., Jiang, Q., Chang, N., Sherbaniuk, R. W., and Taylor, D. E., 1995, Characterization of chromosomal DNA profiles from *Helicobacter pylori* strains isolated from sequential gastric biopsy specimens, *J. Clin. Microbiol.* **33:**2496.
78. Prodinger, W. M., Fille, M., Bauernfeind, A., Stemplinger, I., Amann, S., Pfausler, B., Lass-

Flörl, C., and Dierich, M. P., 1996, Molecular epidemiology of *Klebsiella pneumoniae* producing SHV-5 β-lactamase: Parallel outbreaks due to multiple plasmid transfer, *J. Clin. Microbiol.* **34:**564.

79. Schiappa, D. A., Hayden, M. K., Matushek, M. G., Hashemi, F. N., Sullivan, J., Smith, K. Y., Miyashiro, D., Quinn, J. P., Weinstein, R. A., and Trenholme, G. M., 1996, Ceftazidime-resistant *Klebsiella pneumoniae* and *Escherichia coli* bloodstream infection: A case-control and molecular epidemiologic investigation, *J. Infect. Dist.* **174:**529.
80. Lück, P. C., Helbig, J. H., Hagedorn, H.-J., and Ehret, W., 1995, DNA fingerprinting by pulsed-field gel electrophoresis to investigate a nosocomial pneumonia caused by *Legionella bozemanii* serogroup 1, *Appl. Environ. Microbiol.* **61:**2759.
81. Johnson, W. M., Bernard, K., Marrie, T. J., and Tyler, S. D., 1994, Discriminatory genomic fingerprinting of *Legionella pneumophila* by pulsed-field electrophoresis, *J. Clin. Microbiol.* **32:**2620.
82. Pruckler, J. M., Mermel, L. A., Benson, R. F., Giorgio, C., Cassiday, P. K., Breiman, R. F., Whitney, C. G., and Fields, B. S., 1995, Comparison of *Legionella pneumophila* isolates by arbitrarily primed PCR and pulsed-field gel electrophoresis: Analysis from seven epidemic investigations, *J. Clin. Microbiol.* **33:**2872.
83. Zuerner, R. L., Ellis, W. A., Bolin, C. A., and Montgomery, J. M., 1993, Restriction fragment length polymorphisms distinguish *Leptospira borgpetersenii* serovar hardjo type hardjo-bovis isolates from different geographical locations, *J. Clin. Microbiol.* **31:**578.
84. Herrmann, J. L., Bellenger, E., Perolat, P., Baranton, G., and Girons, I. S., 1992, Pulsed-field gel electrophoresis of *Not*I digests of leptospiral DNA: A new rapid method of serovar identification, *J. Clin. Microbiol.* **30:**1696.
85. Boerlin, P., Bannerman, E., Jemmi, T., and Bille, J., 1996, Subtyping *Listeria monocytogenes* isolates genetically related to the Swiss epidemic clone, *J. Clin. Microbiol.* **34:**2148.
86. Furihata, K, Sato, K., and Matsumoto, H., 1995, Construction of a combined *Not*I/*Sma*I physical and genetic map of *Moraxella (Branhamella) catarrhalis* strain ATCC25238, *Microbiol. Immunol.* **39:**745.
87. Kawakami, Y., Ueno, I., Katsuyama, T., Furihata, K., and Matsumoto, H., 1994, Restriction fragment length polymorphism (RFLP) of genomic DNA of *Moraxella (Branhamella) catarrhalis* isolates in a hospital, *Microbiol. Immunol.* **38:**891.
88. Maloney, S., Welbel, S., Daves, B., Adams, K., Becker, S., Bland, L., Arduino, M., Wallace, R., Jr., Zhang, Y., Buck, G., Risch, P., and Jarvis, W., 1994, *Mycobacterium abscessus* pseudoinfection traced to an automated endoscope washer: Utility of epidemiologic and laboratory investigation, *J. Infect. Dis.* **169:**1166.
89. Wallace, R. J. Jr., Zhang, Y., Brown, B. A., Fraser, V., Mazurek, G. H., and Maloney, S., 1993, DNA large restriction fragment patterns of sporadic and epidemic nosocomial strains of *Mycobacterium chelonae* and *Mycobacterium abscesses, J. Clin. Microbiol.* **31:**2697.
90. Burki, D. R., Bernasconi, C., Bodmer, T., and Telenti, A., 1995, Evaluation of the relatedness of strains of *Mycobacterium avium* using pulsed-field gel electrophoresis, *Eur. J. Clin. Microbiol. Infect. Dis.* **14:**212.
91. Slutsky, A. M., Arbeit, R. D., Barber, T. W., Rich, J., Von Reyn, C. F., Pieciak, W., Barlow, M. A., and Maslow, J. N., 1994, Polyclonal infections due to *Mycobacterium avium* complex in patients with AIDS detected by pulsed-field gel electrophoresis of sequential clinical isolates, *J. Clin. Microbiol.* **32:**1773.
92. Feizabadi, M. M., Robertson, I. D., Cousins, D. V., and Hampson, D. J., 1996, Genomic analysis of *Mycobacterium bovis* and other members of the *Mycobacterium tuberculosis* complex by isoenzyme analysis and pulsed-field gel electrophoresis, *J. Clin. Microbiol.* **34:**1136.
93. Varnerot, A., Clément, F., Gheorghiu, M., and Vincent-Lévy-Frébault, V., 1992, Pulsed-field gel electrophoresis of representatives of *Mycobacterium tuberculosis* and *Mycobacterium bovis* BCG strains, *FEMS Microbiol. Lett.* **98:**155.

94. Yakrus, M. A. and Straus, W. L., 1994, DNA polymorphisms detected in *Mycobacterium haemophilum* by pulsed-field gel electrophoresis, *J. Clin. Microbiol.* **32:**1083.
95. Picardeau, M., Prod'hom, G., Raskine, L., LePennec, M. P., and Vincent, V., 1997, Genotypic characterization of five subspecies of *Mycobacterium kansasii, J. Clin. Microbiol.* **35:**25.
96. Lévy-Frébault, V. V., Thorel, M.-F., Varnerot, A., and Gicquel, B., 1989, DNA polymorphism in *Mycobacterium paratuberculosis,* "wood pigeon mycobacteria," and related mycobacteria analyzed by field inversion gel electrophoresis, *J. Clin. Microbiol.* **27:**2823.
97. Krause, D. C. and Mawn, C. B., 1990, Physical analysis and mapping of the *Mycoplasma pneumoniae* chromosome, *J. Bacteriol.* **172:**4790.
98. Ladefoged, S. A. and Christiansen, G., 1992, Physical and genetic mapping of the genomes of five *Mycoplasma hominis* strains by pulsed-field gel electrophoresis, *J. Bacteriol.* **174:**2199.
99. Ng, L.-K., Carballo, M., and Dillon, J.-A.R., 1995, Differentiation of *Neisseria gonorrhoeae* isolates requiring proline, citrulline, and uracil by plasmid content, serotyping, and pulsed-field gel electrophoresis, *J. Clin. Microbiol.* **33:**1039.
100. Xia, M. S., Roberts, M. C., Whittington, W. L., Holmes, K. K., Knapp, J. S., Dillon, J. A. R., and Wi, T., 1996, *Neisseria gonorrhoeae* with decreased susceptibility to ciprofloxacin: Pulsed-field gel electrophoresis typing of strains from North America, Hawaii, and the Philippines, *Antimicrob. Agents Chemother.* **40:**2439.
101. Blondeau, J. M., Ashton, F. E., Isaacson, M., Yaschuck, Y., Anderson, C., and Ducasse, G., 1995, *Neisseria meningitidis* with decreased susceptibility to penicillin in Saskatchewan, Canada, *J. Clin. Microbiol.* **33:**1784.
102. Yakubu, D. E. and Pennington, T. H. 1995, Epidemiological evaluation of *Neisseria meningitidis* serogroup B by pulsed-field gel electrophoresis, *FEMS Immunol. Med. Microbiol.* **10:**185.
103. Van Dijck, P., Delmée, M., Ezzedine, H., Deplano, A., and Struelens, M. J., 1995, Evaluation of pulsed-field gel electrophoresis and rep-PCR for the epidemiological analysis of *Ochrobactrum anthropi* strains, *Eur. J. Clin. Microbiol. Infect. Dis.* **14:**1099.
104. Blackwood, R. A., Rode, C. K., Read, J. S., Law, I. H., and Bloch, C. A., 1996, Genomic fingerprinting by pulsed-field gel electrophoresis to identify the source of *Pasteurella multocida* sepsis, *Pediatr. Infect. Dis. J.* **15:**831.
105. Hinterberg, K. and Scherf, A., 1994, PFGE: Improved conditions for rapid and high-resolution separation of *Plasmodium falciparum* chromosomes, *Parasitol. Today* **10:**225.
106. Weinberg, G. A., Dykstra, C. C., Durant, P. J., and Cushion, M. T., 1994, Chromosomal localization of 20 genes to five distinct pulsed-field gel karyotypic forms of rat *Pneumocystis carinii, J. Euk. Microbiol.* **41:**117S.
107. Bingen, E., Bonacorsi, S., Rohrlich, P., Duval, M., Lhopital, S., Brahimi, N., Vilmer, E., and Goering, R. V., 1996, Molecular epidemiology provides evidence of genotypic heterogeneity of multidrug-resistant *Pseudomonas aeruginosa* serotype O:12 outbreak isolates from a pediatric hospital, *J. Clin. Microbiol.* **34:**3226.
108. Hla, S. W., Hui, K. P., Tan, W. C., and Ho, B., 1996, Genome macrorestriction analysis of sequential *Pseudomonas aeruginosa* isolates from bronchiectasis patients without cystic fibrosis, *J. Clin. Microbiol.* **34:**575.
109. Eremeeva, M., Balayeva, N., Ignatovich, V., and Raoult, D., 1995, Genomic study of *Rickettsia akari* by pulsed-field gel electrophoresis, *J. Clin. Microbiol.* **33:**3022.
110. Eremeeva, M. E., Roux, V., and Raoult, D., 1993, Determination of genome size and restriction pattern polymorphism of *Rickettsia prowazekii* and *Rickettsia typhi* by pulsed-field gel electrophoresis, *FEMS Microbiol. Lett.* **112:**105.
111. Irie, K., Araki, H., and Oshima, Y., 1991, A new protein kinase, SSP31, modulating the *SMP3* gene-product involved in plasmid maintenance in *Saccharomyces cerevisiae, Gene* **108:**139.
112. Zerva, L., Hollis, R. J., and Pfaller, M. A., 1996, *In vitro* susceptibility testing and DNA typing of *Saccharomyces cerevisiae* clinical isolates, *J. Clin. Microbiol.* **34:**3031.

113. Vaughan-Martini, A., Martini, A., and Cardinali, G., 1993, Electrophoretic karyotyping as a taxonomic tool in the genus *Saccharomyces, Antonie Van Leeuwenhoek* **63:**145.
114. Killalea, D., Ward, L. R., Roberts, D., De Louvois, J., Sufi, F., Stuart, J. M., Wall, P. G., Susman, M., Schwieger, M., Sanderson, P. J., Fisher, I. S. T., Mead, P. S., Gill, O. N., Bartlett, C. L. R., and Rowe, B., 1996, International epidemiological and microbiological study of outbreak of *Salmonella agona* infection from a ready to eat savoury snack .1. England and Wales and the United States, *Br. Med. J.* **313:**1105.
115. Shohat, T., Green, M. S., Merom, D., Gill, O. N., Reisfeld, A., Matas, A., Blau, D., Gal, N., and Slater, P. E., 1996, International epidemiological and microbiological study of outbreak of *Salmonella agona* infection from a ready to eat savoury snack .2. Isreal, *Br. Med. J.* **313:**1107.
116. Baquar, N., Burnens, A., and Stanley, J., 1994, Comparative evaluation of molecular typing of strains from a national epidemic due to *Salmonella brandenburg* by rRNA gene and IS*200* probes and pulsed-field gel electrophoresis, *J. Clin. Microbiol.* **32:**1876.
117. Liebisch, B. and Schwarz, S., 1996, Evaluation and comparison of molecular techniques for epidemiological typing of *Salmonella enterica* subsp *enterica* serovar *dublin, J. Clin. Microbiol.* **34:**641.
118. Navarro, F., Llovet, T., Echeita, M. A., Coll, P., Aladueña, A., Usera, M. A., and Prats, G., 1996, Molecular typing of *Salmonella enterica* serovar typhi, *J. Clin. Microbiol.* **34:**2831.
119. Murase, T., Nakamura, A., Matsushima, A., and Yamai, S., 1996, An epidemiological study of *Salmonella enteritidis* by pulsed-field gel electrophoresis (PFGE): Several PFGE patterns observed in isolates from a food poisoning outbreak, *Microbiol. Immunol.* **40:**873.
120. Ridley, A. M., Punia, P., Ward, L. R., Rowe, B., and Threlfall, E. J., 1996, Plasmid characterization and pulsed-field electrophoretic analysis demonstrate that ampicillin-resistant strains of *Salmonella enteritidis* phage type 6a are derived from *Salmonella enteritidis* phage type 4, *J. Appl. Bacteriol.* **81:**613.
121. Thong, K.-L., Puthucheary, S., Yassin, R. M., Sudarmono, P., Padmidewi, M., Soewandojo, E., Handojo, I., Sarasombath, S., and Pang, J., 1995, Analysis of *Salmonella typhi* isolates from Southeast Asia by pulsed-field gel electrophoresis, *J. Clin. Microbiol.* **33:**1938.
122. Thong, K. L., Passey, M., Clegg, A., Combs, B. G., Yassin, R. M., and Pang, T., 1996, Molecular analysis of isolates of *Salmonella typhi* obtained from patients with fatal and nonfatal typhoid fever, *J. Clin. Microbiol.* **34:**1029.
123. Liu, S.-L., Hessel, A., and Sanderson, K. E., 1993, The *Xba*I-*Bln*I-*Ceu*I genomic cleavage map of *Salmonella typhimurium* LT2 determined by double digestion, end labelling, and pulsed-field gel electrophoresis, *J. Bacteriol.* **175:**4104.
124. Murase, T., Okitsu, T., Suzuki, R., Morozumi, H., Matsushima, A., Nakamura, A., and Yamai, S., 1995, Evaluation of DNA fingerprinting by PFGE as an epidemiologic tool for *Salmonella* infections, *Microbiol. Immunol.* **39:**673.
125. Miranda, G., Kelly, C., Solorzano, F., Leanos, B., Coria, R., and Patterson, J. E., 1996, Use of pulsed-field gel electrophoresis typing to study an outbreak of infection due to *Serratia marcescens* in a neonatal intensive care unit, *J. Clin. Microbiol.* **34:**3138.
126. Shi, Z. Y., Liu, P. Y. F., Lau, Y. J., Lin, Y. H., and Hu, B. S., 1997, Use of pulsed-field gel electrophoresis to investigate an outbreak of *Serratia marcescens, J. Clin. Microbiol.* **35:**325.
127. Sader, H. S., Peri, T. M., Hollis, R. J., Divishek, D., Herwaldt, L. A., and Jones, R. N., 1994, Nosocomial transmission of *Serratia odorifera* biogroup 2: Case report demonstration by macrorestriction analysis of chromosomal DNA using pulsed-field gel electrophoresis, *Infect. Control Hosp. Epidemiol.* **15:**390.
128. Kariuki, S., Muthotho, N., Kimari, J., Waiyaki, P., Hart, C. A., and Gilks, C. F., 1996, Molecular typing of muti-drug resistant *Shigella dysenteriae* type 1 by plasmid analysis and pulsed-field gel electrophoresis, *Trans. R. Soc. Trop. Med. Hyg.* **90:**712.
129. Rajakumar, K., Sasakawa, C., and Adler, B., 1996, Spontaneous 99-kb chromosomal deletion

results in multi-antibiotic susceptibility and an attentuation of contact haemolysis in *Shigella flexneri* 2a, *J. Med. Microbiol.* **45:**64.

130. Brian, M. J., Van, R., Townsend, I., Murray, B. E., Cleary, T. G., and Pickering, L. K., 1993, Evaluation of the molecular epidemiology of an outbreak of multiply resistant *Shigella sonnei* in a day-care center by using pulsed-field gel electrophoresis and plasmid DNA analysis, *J. Clin. Microbiol.* **31:**2152.
131. Liu, P.Y.-F., Lau, Y.-J., Hu, B.-S., Shyr, J. M., Shi, Z.-Y., Tsai, W.-S., Linn, Y.-H., and Tseng, C.-Y., 1995, Analysis of clonal relationships among isolates of *Shigella sonnei* by different molecular typing methods, *J. Clin. Microbiol.* **33:**1779.
132. Tateishi, T., Murayama, S. Y., Otsuka, F., and Yamaguchi, H., 1996, Karyotyping by PFGE of clinical isolates of *Sporothrix schenckii, FEMS Immunol. Med. Microbiol.* **13:**147.
133. Branger, C., Gardye, C., and Lambert-Zechovsky, N., 1996, Persistence of *Staphylococcus aureus* strains among cystic fibrosis patients over extended periods of time, *J. Med. Microbiol.* **45:**294.
134. Liu, P. Y. F., Shi, Z. Y., Lan, Y. J., Hu, B. S., Shyr, J. M., Tsai, W. S., Lin, Y. H., and Tseng, C. Y., 1996, Use of restriction endonuclease analysis of plasmids and pulsed-field gel electrophoresis to investigate outbreaks of methicillin-resistant *Staphylococcus aureus* infection, *Clin. Infect. Dis.* **22:**86.
135. Huebner, J., Pier, G. B., Maslow, J. N., Muller, E., Shiro, H., Parent, M., Kropec, A., Arbeit, R. D., and Goldmann, D. A., 1994, Endemic nosocomial transmission of *Staphylococcus epidermidis* bacteremia isolates in a neonatal intensive care unit over 10 years, *J. Infect. Dis.* **169:**526.
136. Lina, B., Forey, F., Tigaud, J. D., and Fleurette, J., 1995, Chronic bacteraemia due to *Staphylococcus epidermidis* after bone marrow transplantation, *J. Med. Microbiol.* **42:**156.
137. Degener, J. E., Heck, M. E. O. C., Van Leeuwen, W. J., Heemskerk, C., Crielaard, A., Joosten, P., and Caesar, P., 1994, Nosocomial infection by *Staphylococcus haemolyticus* and typing method for epidemiological study, *J. Clin. Microbiol.* **32:**2260.
138. Breen, J. D. and Karchmer, A. W., 1994, Usefulness of pulsed-field gel electrophoresis in confirming endocarditis due to *Staphylococcus lugdunensis, Clin. Infect. Dis.* **19:**985.
139. Laing, F. P. Y., Ramotar, K., Read, R. R., Alfieri, N., Kureishi, A., Henderson, E. A., and Louie, T. J., 1995, Molecular epidemiology of *Xanthomonas maltophilia* colonization and infection in the hospital environment, *J. Clin. Microbiol.* **33:**513.
140. Ramage, L., Green, K., Pyskir, D., and Simor, A. E., 1996, An outbreak of fatal nosocomial infections due to a group A streptococcus on a medical ward, *Infect. Control Hosp. Epidemiol.* **17:**429.
141. Stanley, J., Desai, M., Xerry, J., Tanna, A., Efstratiou, A., and George, R., 1996, High-resolution genotyping elucidates the epidemiology of group A streptococcus outbreaks, *J. Infect. Dis.* **174:**500.
142. Fasola, E., Livdahl, C., and Ferrieri, P., 1993, Molecular analysis of multiple isolates of the major serotypes of group B streptococci, *J. Clin. Microbiol.* **31:**2616.
143. Gordillo, M. E., Singh, K. V., Baker, C. J., and Murray, B. E., 1993, Typing of group B streptococci: Comparison of pulsed-field gel electrophoresis and conventional electrophoresis, *J. Clin. Microbiol.* **31:**1430.
144. Hall, L. M. C., Whiley, R. A., Duke, B., George, R. C., and Efstratiou, A., 1996, Genetic relatedness within and between serotypes of *Streptococcus pneumoniae* from the United Kingdom: Analysis of multilocus enzyme electrophoresis, pulsed-field gel electrophoresis, and antimicrobial resistance patterns, *J. Clin. Microbiol.* **34:**853.
145. Moissenet, D., Valcin, M., Marchand, V., Garabédian, E. N., Geslin, P., Garbarg-Chenon, A., and Vu-Thien, H., 1997, Molecular epidemiology of *Streptococcus pneumoniae* with decreased susceptibility to penicillin in a Paris Children's Hospital, *J. Clin. Microbiol.* **35:**298.
146. Gonzalez, J., Muñoz, S., Ortiz, S., Anacona, D., Salgado, S., Galleguillos, M., Neira, II., Sagua,

H., and Solari, A., 1995, Biochemical, immunological, and biological characterization of *Trypanosoma cruzi* populations of the Andean north of Chile, *Exp. Parasitol.* **81:**125.

147. Henriksson, J., Solari, A., Rydker, M., Sousa, O. E., and Pettersson, U., 1996, Karyotype variability in *Trypanosoma rangeli, Parasitology* **112:**385.

148. Skov, M. N., Pedersen, K., and Larsen, J. L., 1995, Comparison of pulsed-field gel electrophoresis, ribotyping, and plasmid profiling for typing of *Vibrio anguillarum* serovar O1, *Appl. Environ. Microbiol.* **61:**1540.

149. Evins, G. M., Cameron, D. N., Wells, J. G., Greene, K. D., Popovic, T., Giono-Cerezo, S., Wachsmuth, I. K., and Tauxe, R. V., 1995, The emerging diversity of the electrophoretic types of *Vibrio cholerae* in the Western Hemisphere, *J. Infect. Dis.* **172:**173.

150. Popovic, T., Fields, P. I., Olsvik, O., Wells, J. G., Evins, G. M., Cameron, D. M., Farmer, J. J. III, Bopp, C. A., Wachsmuth, K., Sack, R. B., Albert, M. J., Nair, G. B., Shimada, T., and Feeley, J. C., 1995, Molecular subtyping of toxigenic *Vibrio cholerae* O139 causing epidemic cholera in India and Bangladesh, 1992–1993, *J. Infect. Dis.* **171:**122.

151. Wong, H. C., Lu, K. T., Pan, T. M., Lee, C. L., and Shih, D. Y. C., 1996, Subspecies typing of *Vibrio parahaemolyticus* by pulsed-field gel electrophoresis, *J. Clin. Microbiol.* **34:**1535.

152. Buchrieser, C., Gangar, V. V., Murphree, R. L., Tamplin, M. L., and Kaspar, C. W., 1995, Multiple *Vibrio vulnificus* strains in oysters as demonstrated by clamped homogeneous electric field gel electrophoresis, *App. Environ. Microbiol.* **61:**1163.

153. Tamplin, M. L., Jackson, J. K., Buchrieser, C., Murphree, R. L., Portier, K. M., Gangar, V., Miller, L. G., and Kaspar, C. W., 1996, Pulsed-field gel electrophoresis and ribotype profiles of clinical and environmental *Vibrio vulnificus* isolates, *Appl. Environ. Microbiol.* **62:**3572.

154. Iteman, I., Guiyoule, A., and Carniel, E., 1996, Comparison of three molecular methods for typing and subtyping pathogenic *Yersinia enterocolitica* strains, *J. Med. Microbiol.* **45:**48.

155. Najdenski, H., Iteman, I., and Carniel, E., 1994, Efficient subtyping of pathogenic *Yersinia enterocolitica* strains by pulsed-field gel electrophoresis, *J. Clin. Microbiol.* **32:**2913.

156. Lucier, T. S., and Brubaker, R. R., 1992, Determination of genome size, macrorestriction pattern polymorphism, and nonpigmentation-specific deletion in *Yersinia pestis* by pulsed-field gel electrophoresis, *J. Bacteriol.* **174:**2078.

157. Tenover, F. C., Arbeit, R. D., Goering, R. V., Mickelsen, P. A., Murray, B. E., Persing, D. H., and Swaminathan, B., 1995, Interpreting chromosomal DNA restriction patterns produced by pulsed-field gel electrophoresis: Criteria for bacterial strain typing, *J. Clin. Microbiol.* **33:**2233.

158. Correia, A., Martín, J. F., and Castro, J. M., 1994, Pulsed-field gel electrophoresis analysis of the genome of amino acid producing corynebacteria: Chromosome sizes and diversity of restriction patterns, *J. Gen. Microbiol.* **140:**2841.

159. Goering, R. V., 1995, The application of pulsed-field gel electrophoresis to the analysis of global dissemination of methicillin-resistant *Staphylococcus aureus,* in: *Maurice Rapin Colloquia: Methicillin Resistant Staphylococci* (C. Brun-Buisson, M. W., Casewell, N. El Solh and B. Régnier, eds.), Flammarion Médecine-Sciences; Paris, pp. 75–81.

160. Barrett, T. J., Lior, H., Green, J. H., Khakhria, R., Wells, J. G., Bell, B. P., Greene, K. D., Lewis, J., and Griffin, P. M., 1994, Laboratory investigation of a multistate food borne outbreak of *Escherichia coli* O157:H7 by using pulsed-field gel electrophoresis and phage typing, *J. Clin. Microbiol.* **32:**3013.

161. Borecká, P., Rosypal, S., Pantucek, R., and Doskar, J., 1996, Localization of prophages of serological group B and F on restriction fragments defined in the restriction map of *Staphylococcus aureus* NCTC 8325, *FEMS Microbiol. Lett.* **143:**203.

162. Lina, B., Bes, M., Vandenesch, F., Greenland, T., Etienne, J., and Fleurette, J., 1993, Role of bacteriophages in genomic variability of related coagulase-negative staphylococci, *FEMS Microbiol. Lett.* **109:**273.

163. Perkins, J. D., Heath, J. D., Sharma, B. R., and Weinstock, G. M., 1993, *Xba*I and *Bln*I genomic

cleavage maps of *Escherichia coli* K-12 strain MG1655 and comparative analysis of other strains, J. Mol. Biol. **232:**419.

164. Rodley, P. D., Römling, U., and Tümmler, B., 1995, A physical genome map of the *Burkholderia cepacia* type strain, *Mol. Microbiol.* **17:**57.
165. Iandolo, J. J., Bannantine, J. P,. and Stewart, G. C., 1997, Genetic and physical map of the chromosome of *Staphylococcus aureus*, in: *The Staphylococci in Human Disease* (K. B. Crossley and G. L. Archer, eds.), Churchill Livingstone, New York, pp. 39–53.
166. Winters, M. A., Goering, R. V., Boon, S. E., Morin, R., Sorensen, M., and Snyder, L., 1993, Epidemiological analysis of methicillin-resistant *Staphylococcus aureus* comparing plasmid typing with chromosomal analysis by field inversion gel electrophoresis, *Med. Microbiol. Lett.* **2:**33.

10

Tyramide Signal Amplification

Applications in Detecting Infectious Agents

G. J. LITT and M. N. BOBROW

1. INTRODUCTION

→ See P 171

In 1989, we reported a new technique[1] for signal amplification which we originally termed catalyzed reporter deposition or CARD. The version discussed in this paper is more properly termed tyramide signal amplification (TSA) because tyramine is used as the reactive part of the substrate. TSA, CARD, catalyzed signal amplification (CSA) and biotin signal amplification are all terms that have been used in the literature. The methodology was described for a chromogenic, indirect format in microplate immunoassays (ELISA) and several of our examples included the sensitive detection of infectious agents, such as herpes simplex virus (HSV) and the p24 core protein of the human immunodeficiency virus type 1 (HIV-1). The technique is sensitive, yielding ELISA amplifications of 15–30-fold with significant improvements in signal-to-noise and also has unusual flexibility and simplicity. In our subsequent description of use in Western blotting[2], we noted another interesting characteristic, the virtually complete absence of band diffusion, and we attributed this to the short half-life of a free radical intermediate. Since that time, a number of reports have appeared that have validated our original work and also have greatly extended the methodological adaptations and

G. J. LITT and M. N. BOBROW • Dupont-NEN Life Science Products, Medical Products Department, Boston, Massachusetts 02118.

Rapid Detection of Infectious Agents, edited by Specter *et al.* Plenum Press, New York, 1998.

analyte applications in both direct and indirect modes for the high-sensitivity assay of a variety of proteins and nucleic acids.

This presentation reviews the technique and its variants, focuses on the utility to date for the detecting low levels of infectious disease analytes, and finally outlines our view of the future utilization of the technology in this important area of research and diagnosis.

2. BACKGROUND AND TERMINOLOGY

There have been many descriptions of nonradiometric signal amplification techniques described during the last 10–15 years, and we present here only a brief overview of some of the reviews and more interesting contributions. Ngo[3] and Kricka[4] have reviewed high-sensitivity immunoassay techniques, ranging from novel sensor devices to the use of amplifying and multiple labels, a variety of enzyme cascade formats, and the application of luminescent detection. Diamandis and Christopoulos[5] contributed a comprehensive examination of the extremely powerful and widely used biotin-(strept)avidin system. van Gijlswijk *et al.*[6] described a variant on this format called "shuttles" where avidin-labeled antibodies increase sensitivity five- to eightfold and Grumbach and Veh[7] showed that improved avidin biotin complexes (ABC) yield about a three- to eightfold improvement in ELISA and a considerable (not quantified) amplification in immunocytochemical staining.

For nucleic acid systems, target rather than signal amplification has attracted the most interest. The polymerase chain reaction[8] (PCR) has been applied and validated for high-sensitivity detection of a wide variety of infectious agents although quantitation is quite complicated to achieve and issues of diffusion cloud some of the *in situ* applications. An interesting variant is the use of immuno-PCR[9] where the detector antibody used in a typical solid-phase sandwich ELISA is labeled with a stretch of single-stranded DNA that is amplified by PCR and detected conventionally.

Finally, Larison *et al.*[10] described a novel fluorogenic alkaline phosphatase substrate for immunohistochemistry applications that is precipitated in the microenvironment of the enzyme, yielding a highly photostable sensitive signal.

3. PRINCIPLES

3.1. Basis

CARD is a signal amplification technique based on using the enzyme immobilized in the initial steps of detecting an analyte to convert or activate a labeled substrate to a form that results in covalent bonding of the activated moiety to the immediately adjacent surface. In a very short time, multiple deposition of the labeled substrate occurs, and subsequent detection yields an effectively large amplification. The system is essentially independent of the protocol, targets, or methodology used before amplification. The initiating enzyme can be from primary, secondary or multiplicative detector formats and may, for example, be part of a sandwich immunoassay or an indirect detector in a nucleic acid hybridization format.

In tyramide signal amplification (TSA), the enzyme is horseradish peroxidase (HRP), and the substrate comprises a label (originally biotin or fluorescein) covalently coupled to tyramine. The enzymic activation, oxidation in this case, results in the formation of a free radical in a position ortho to the phenolic group in the tyramide. Under normal solution phase conditions, the predominant result of the oxidation is the formation of a dityramide, and the radicalized substrate reacts with an unconverted molecule. The mechanism for this reaction and the subsequent dimeric product(s) was originally proposed by Gross and Sizer[11] as shown in Fig. 1. Their work showed that all the substrates they tested with free phenolic groups were oxidized by HRP to form dimers, notably including the amino acid tyrosine. This reaction was the basis for the development of fluorogenic HRP substrates by Guilbault *et al.*[12,13] and by Zaitsu and Ohkura[14].

We conceptualized[1,2] that in situations where a solid phase containing po-

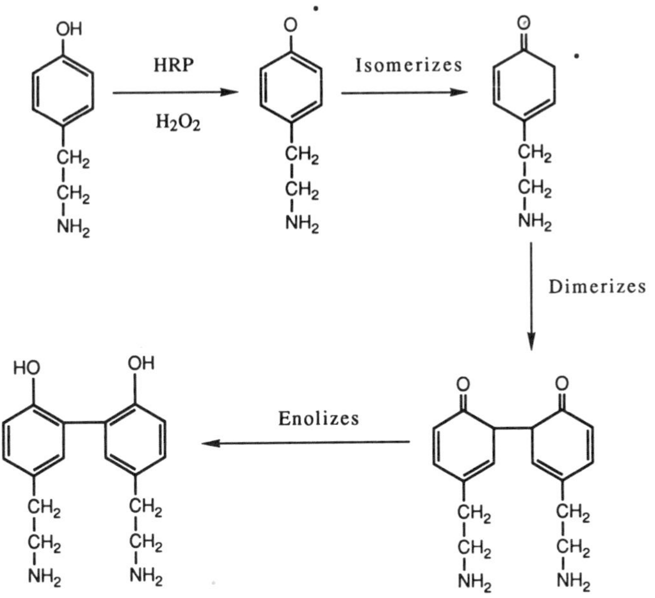

FIGURE 1. Proposed mechanism for dityramide formation by peroxidase oxidation.[11]

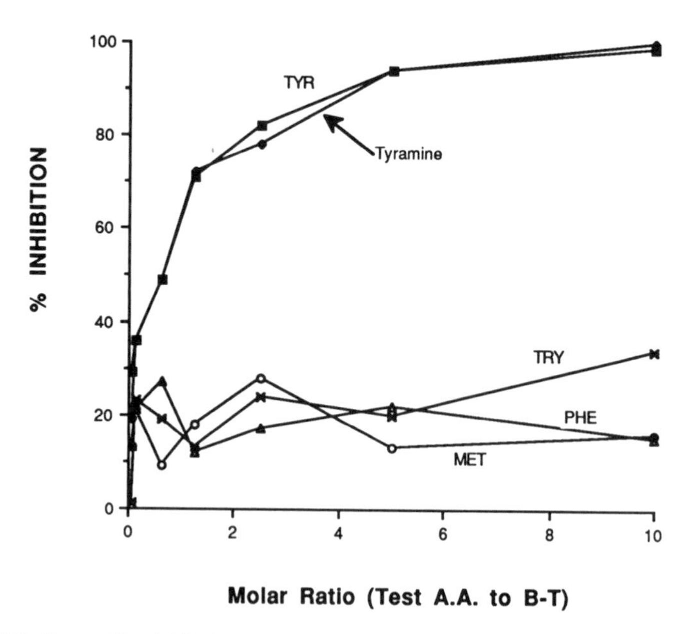

FIGURE 2. Competitive inhibition of biotinyl tyramide by various amino acids (and a tyramine control).

tential reactive couplers is available (or could be added) in the microenvironment, appropriate changes in reaction conditions favor the ratio of activated labeled substrate that covalently links to the immobilized coupler (rather than forming a dimer). In TSA, this is facilitated by (1) assuring the local availability of such immobilized couplers and (2) modulating the concentration of the labeled substrate to reduce the probability of solution phase dimer formation. Our original postulate in ELISA and blotting applications was that the commonly used "blocking" reagents that prevent nonspecific binding to the surface either already contain potentially reactive couplers for the activated free radical or can be chemically modified with couplers. Figure 2 shows an experiment to pinpoint the reactive moiety. In this study, the ability of the most likely amino acids to inhibit the enzyme-mediated deposition was tested in a microplate ELISA format. It is quite clear that tyrosine is the most probable reactant because it was the only material tested that compared closely with the tyramine control. The probable mechanism is shown schematically in Fig. 3. As noted above, Gross and Sizer[11] showed earlier that tyrosine is a suitable substrate for HRP oxidation and this amino acid is a component of the most commonly used blocking reagents, such as bovine serum albumin (BSA), casein, and the various formulations of nonfat dry milk.

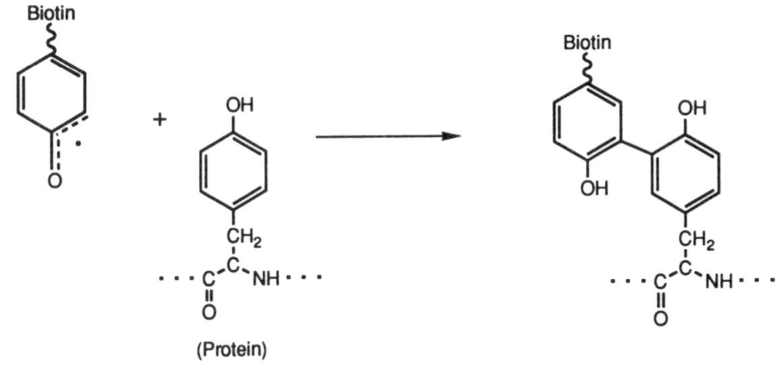

FIGURE 3. Schematic reactivity of peroxidase-activated tyramide substrate with tyrosine portion of immobilized protein.

In histochemical and *in situ* hybridization applications of TSA, the tissue protein itself is adequate, and it is usually not necessary to add an exogenous protein. The short half-life of the free radical produced by the oxidative step and the availability of reactive "receptors" in the microenvironment (the endogenous protein matrix of the sample) also minimize diffusion and result in morphologically high fidelity of the reaction product's location.

3.2. Formats

3.2.1. *Indirect*

Our original publications[1,2] emphasized using a biotinyl tyramide substrate in an indirect mode with labeled streptavidin detection. Adams[15] showed that this format had great versatility for lectin histochemistry and immunohistochemical (IHC) staining and suggested that high-sensitivity versions would be valuable for *in situ* hybridization where biotin labeled probes are employed. Although extra detection and wash steps are required in this configuration, great flexibility is possible and often, but not always, higher overall amplification is obtained. A number of alternative hapten-containing substrates have been synthesized and paired with appropriate binders containing a variety of labels. Table I lists most of the combinations being used.

One area of flexibility of unusual interest is the ability to switch the final detector label. This was clearly demonstrated in the original publications[1,2] and was emphasized by Adams[15] for histochemical applications. As one example, biotinyl- or fluorescein tyramide is deposited from activation by the original HRP label, and this can be followed up with either streptavidin- or antifluorescein-

TABLE I
Indirect TSA Systems

Substrate	Binders	Label
Biotinyl tyramide (B-T)	Streptavidin, avidin	Enzyme (e.g., HRP, AP)
Biotinyl tyramide	Streptavidin, avidin	Fluor (e.g., Fluorescein, Texas Red®, 7-hydroxycoumarin)
Fluorescein tyramide (F-T)	Antifluorescein antibody	Enzyme (as B-T)
Fluorescein tyramide	Antifluorescein antibody	Fluor (as B-T)

alkaline phosphatase. The final detection could be done with, for example, a chromogenic alkaline phosphatase substrate, such as BCIP/NBT.

3.2.2. *Direct*

In recent years, the rapid spread of TSA technology to microscopical methods, such as immunofluorescence (IF) and fluorescent *in situ* hybridization (FISH), has brought with it the potential for cutting down on secondary manipulations by employing directly detectable fluorescent substrates. The demonstrated amplification available with TSA has shown that adequate sensitivity is usually attainable and simplified formats are possible. Because fluorescein tyramide had already been shown to be a functional conjugate (although originally used as part of a ligand-receptor pair with labeled antifluorescein), it has served as the model for synthesizing a variety of fluor-containing substrates. Table II, adapted from the pending comprehensive report of van Gijlswijk *et al.*[16], lists the fluorescent tyramide derivatives tested by these authors. A series of useful comparisons of the various compounds for IHC, immuno-cytochemistry (ICC), FISH, and multicolor applications are to be found in this paper. Although these authors had poor success with Texas Red® tyramide as a substrate, the use of this fluor as part of a secondary detector (e.g., linked to streptavidin) has proven very successful.

TABLE II
Fluorescent Tyramide Derivatives[a]

Aminomethylcoumarin	Rhodamine Green™	X-Rhodamine
7-Hydroxycoumarin	Eosin	Texas Red®
Fluorescein	Rhodamine	Cy3™
BODIPY® FL	Rhodamine Red™	SI-Red

[a]Abstracted from the comprehensive tabulation of von Gijlswijk *et al.*[16]

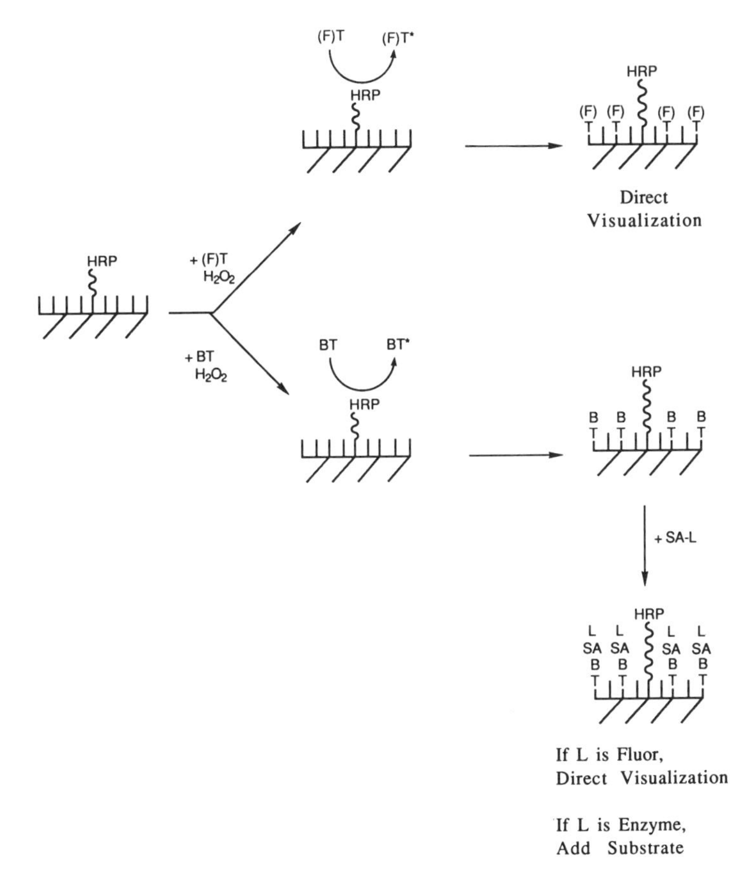

FIGURE 4. Schematic of the indirect and direct versions of TSA.

Figure 4 schematically shows both the indirect and direct formats of TSA as applied to immunohistochemistry and *in situ* hybridization.

4. APPLICATIONS

One of the more important surrogate markers for HIV-1 infection is the relatively immunogenic protein designated as p24 that makes up a large part of the core of the virus. The HIV-1 p24 immunoassay is still widely used as a prognostic aid by members of the AIDS Clinical Therapeutics Group (ACTG)[17] and others to improve the sensitivity of peripheral blood mononuclear cell coculture methodology. These data are particularly important to monitor viral load

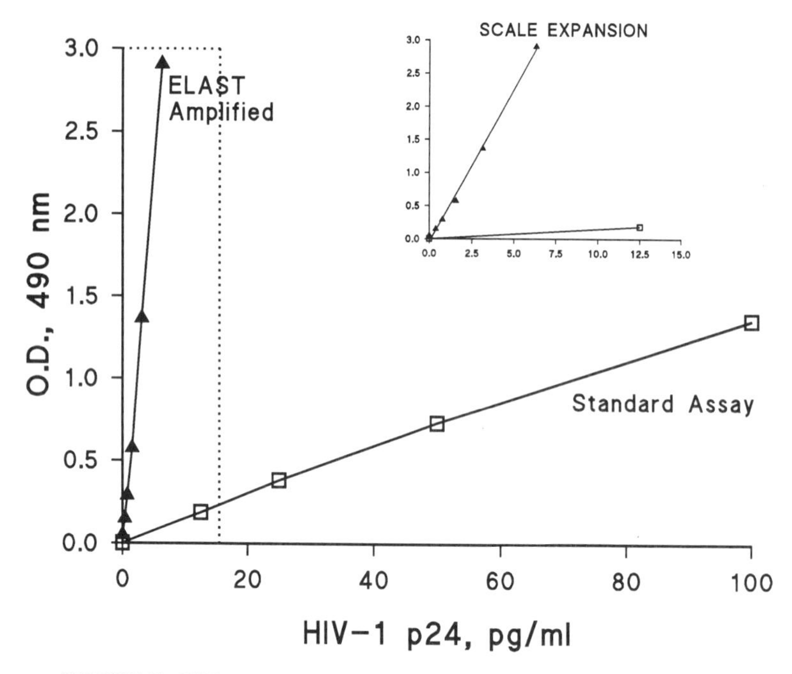

FIGURE 5. HIV-1 p24 standard curves with and without TSA amplification.

and assure prompt initiation and/or monitoring of drug therapy. Replacing the previous radiometric reverse transcriptase assay by immunoassay led to a significant improvement in precision and cut assay time and manipulations considerably. In recent years, the immunoassay has been extended to detecting p24 in serum and plasma although initial results were quite inaccurate because the antigen was masked by endogenous antibody.

In 1989, in the seminal paper on CARD/TSA, Bobrow *et al.*,[1] described amplification of greater than eightfold in the HIV-1 p24 assay while retaining acceptable levels of precision. Figure 5 is an example of the increased analytical sensitivity. Results of this type led to the commercialization by DuPont/NEN Life Science Products of the amplification system dedicated to ELISA (named ELAST for ELISA Amplification SysTem). As refinements of the HIV-1 p24 system proceeded, augmentation with the ELAST system significantly improved clinical sensitivity in serum and plasma samples where the immune complexes were disrupted by acid or heat treatment.

In an unpublished study at NEN Life Science Products[18], the ELAST-augmented HIV-1 p24 assay was compared to the FDA Licensed Abbott HIV-1 Antigen kit to determine if the extra sensitivity would shorten the time to positivity when testing culture supernatants. In testing 296 samples, the amplified system detected more positive supernatants (56%) after seven days than the reference Abbott test did at 14 days (48%). In a related portion of the study, it was shown that the ELAST amplified test enabled the qualitative detection of p24 in supernatants from peripheral blood mononuclear cells one to three days earlier than the Abbott test and two days earlier compared with plasma macroculture.

Although many AIDS studies are moving from the HIV-1 p24 surrogate to measuring low levels of RNA using a variant of the PCR, the extra steps, cost, and potential for cross-contamination involved are factors that have encouraged some workers to evaluate the inherently simpler signal amplification format. Recently, Lyamuya *et al.*[19] and Schupbach *et al.*[20] have reported on their successes with the ELAST system combined with a heat-disruption protocol employed for assaying AIDS patient samples. In the former case, the authors concluded that the composite assay ". . . . may be sufficient for the early diagnosis of HIV-1 infection in infants in settings with limited laboratory facilities" and would be useful for predicting the risk of mother-to-infant transmission. Schupbach *et al.*[20] found that the addition of the ELAST steps resulted in a sensitivity equal to the PCR kit for RNA ". . . . but at a fraction of the cost." These authors described a prospective and retrospective blind study of samples from 254 patients where they found that confirmed positives in plasma were 97.8% with the "boosted ELISA" compared with 95.7% with RNA PCR. It is interesting to note that the high sensitivity of the ELAST format helped significantly in making the heat immune complex disruption practical. Coagulation problems could be largely avoided by dilution.

The ELAST system also improves sensitivity in detecting herpes simplex viral antigen by ELISA. Bobrow *et al.*[1] showed a 16-fold improvement in analytical sensitivity with this assay. In addition, utilizing the amplification procedure for Western blotting of HSV antigen on nitrocellulose, Bobrow and his colleagues[2] obtained sensitivity increases of 16- to 32-fold, using 4-chloronapthol as a substrate, and eightfold with diaminobenzidine.

In contrast to the results described, Deelder *et al.*[21] found that a modified ELISA method incorporating TSA did not ". . . result in assays that were consistently better than the 'standard' assay . . ." for the circulating anodic antigen (CAA) of schistosoma.

An example of the utility of TSA for immunohistochemistry, illustrated by a comparison done at NEN Life Science Products is shown in Figs. 6(A) and (B). In Figure 6(A), cytomegalovirus (CMV)-infected MRC-5 cells were detected by standard IHC using CY3-labeled early-immediate-region mouse monoclonal antibody at a suboptimal dilution. In contrast, the TSA-enhanced version (primary detection with the same low concentration of unlabeled antibody followed by antimouse-HRP and tetramethylrhodamine tyramide) is shown in Figure 6(B). Although admittedly an artificial comparison due to the use of a low concentra-

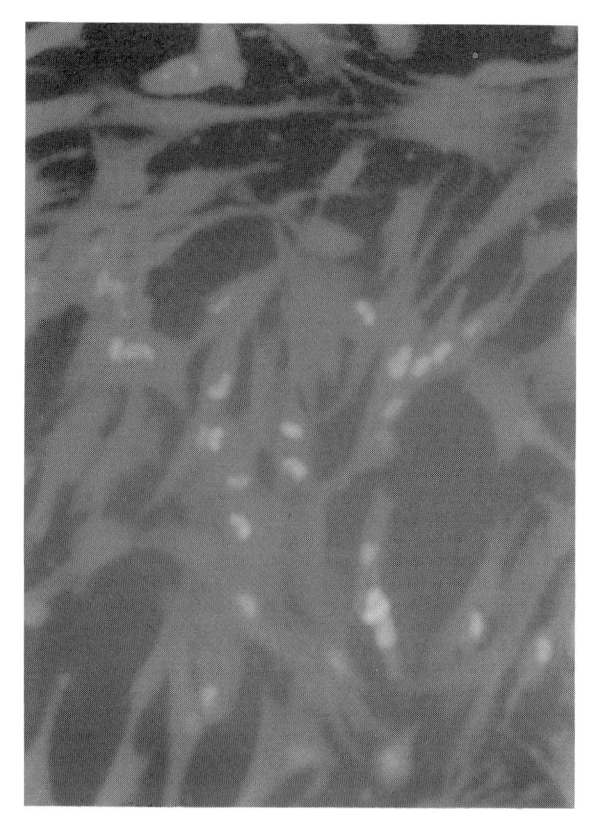

FIGURE 6. (A) CV-infected MRC-5 cells detected by standard IHC using (a suboptiomal level of) CY3-labeled early-immediate-region mouse monoclonal antibody. (B) Same antibody concentration enhanced with indirect TSA. See text for more details.

tion of primary antibody, the relative improvements in sensitivity, retention of localization, and lack of background are notable.

Luka and Afflerbach[22] evaluated direct TSA using fluorescein tyramide for detecting low copy number viral genomes in both lymphocytes and tissues. For proviral HIV DNA in CEM/RK cells, between five and eight copies could be counted as well defined areas within the cells using TSA whereas *in situ* PCR of parallel samples yielded similar sensitivity but resulted in diffuse staining over the entire cell. In addition, they found that fewer false negatives were seen in these sample after TSA than with *in situ* PCR. These authors also studied the use of the amplification system for low copy sequences of proviral HIV in biopsy and autopsy material. In samples of this type, autofluorescence is often a problem and this proved true with the direct fluorescent amplification. However, the authors

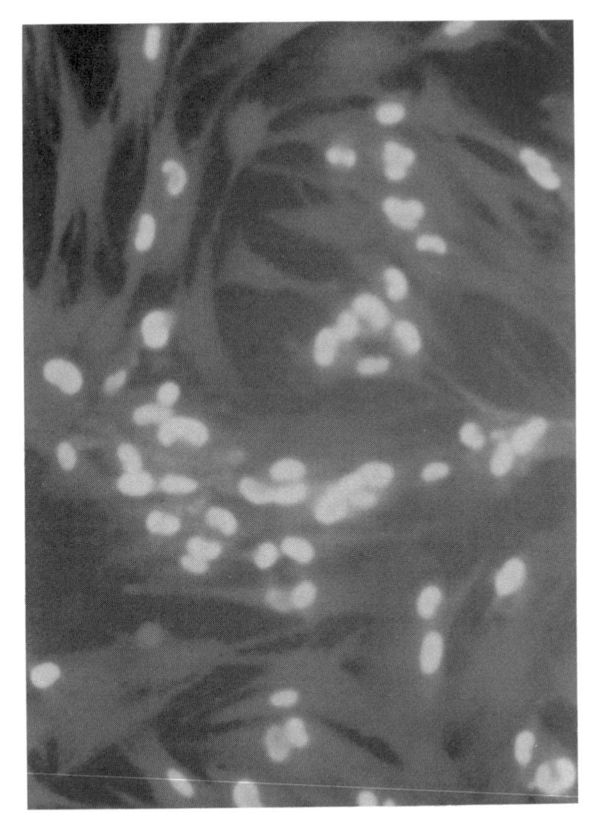

FIGURE 6. *(Continued).*

took advantage of the ability to switch detector and visualization. In this case, they used fluorescein tyramide to deposit but instead of direct viewing, added the steps of detection with antifluorescein-alkaline phosphatase followed by chromogenic visualization with BCIP/NBT. It was found that the signals now visible in the light microscope yielded even further amplification and they detected proviral DNA in various tissues using only the *gag* region for detection. Their work with lymphoid tissue samples confirmed the comparative sensitivity and the localization advantage. With TSA, a strong signal was observed with deposits localized close to the viral copies in the nucleus whereas *in situ* PCR also yielded a strong signal but the staining was distributed over the entire cell.

Another example of the utility of TSA to detect low copy numbers of infectious agent genes is shown in Figures 7(A) and (B), the work of Karl Adler at NEN Life Science Products. In this example, human papillomavirus (HPV)-16 was detected in SiHa cells [Figure 7(B)] by the indirect format with a biotinylated

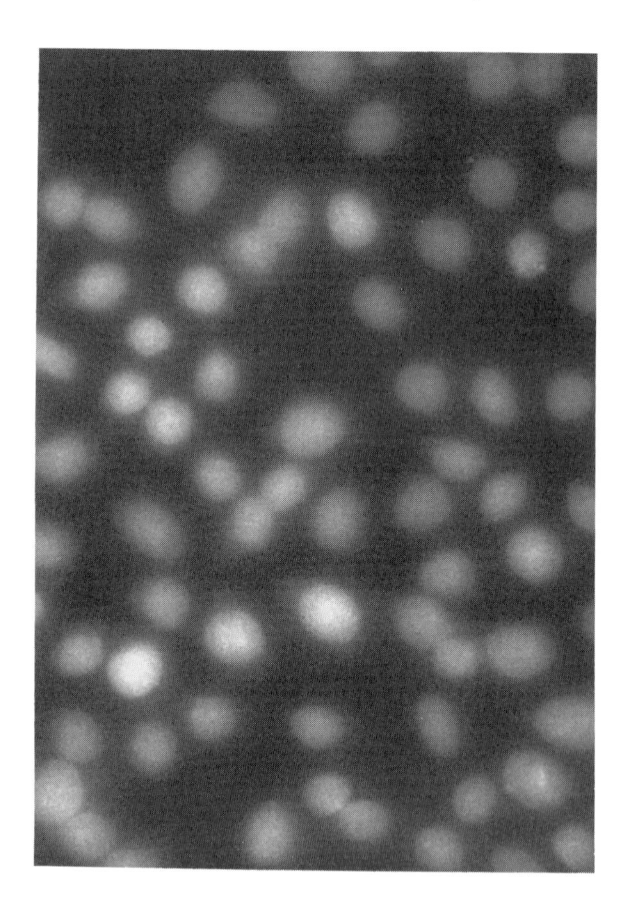

FIGURE 7. HPV-16 detection. (A) Detroit 511 cells counterstained with Hoechst Dye 33342. (B) HPV-16 in similarly counterstained SiHa cells (two copies/cell) detected with indirect TSA using streptavidin-Texas Red™. More details are in the text.

probe and detection with streptavidin-Texas Red™. Nuclei were counterstained with Hoechst Dye 33342. For comparison, Fig. 7(A) shows Detroit 511, a negative cell line.

In a study designed to optimize the use of biotinylated oligonucleotide probes, Zecchini *et al.*[23] attained a detection of less that 0.02 fmol of final *Clostridium tyrobutyricum* DNA using a dot blot assay. These researchers used the indirect TSA system with biotinyl tyramide and Neutralite© avidin-peroxidase and found a 10-fold improvement in sensitivity with deeper intensity of color, no noticeable elevation in background, and insignificant increases in assay time.

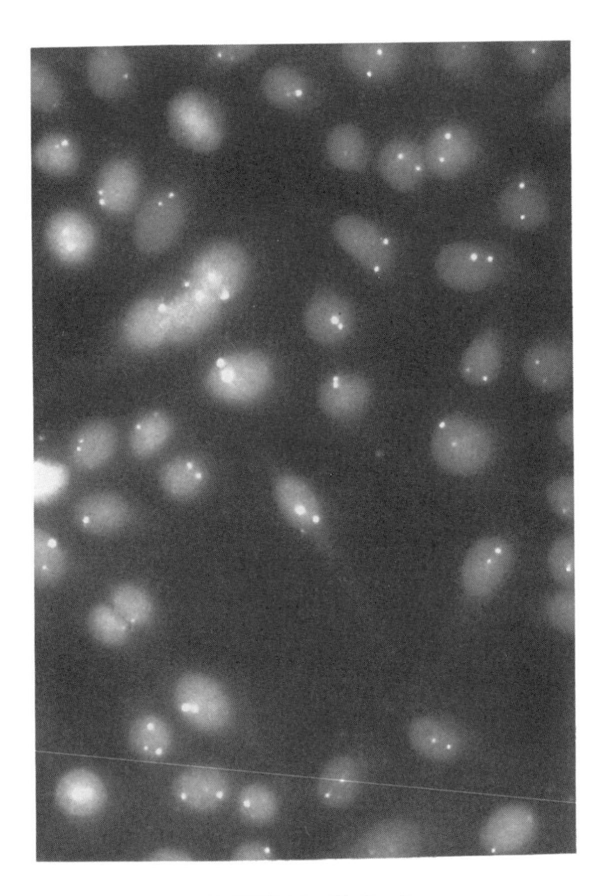

FIGURE 7. *(Continued).*

5. DISCUSSION

The TSA methodology was originally developed to increase signals in solid-phase immunoassays and in Western blotting where the primary proof examples were such infectious disease analytes as the p24 core protein of HIV-1, HSV antigen, and CMV antigen. Although our review of ELISA takes up a large part of the previous application discussion, there has been a major shift in recent years. The short half-life of the activated substrate combined with the unique specificity made possible by covalent deposition has led to a surge in applications for detection and localization in cytology, immunohistochemistry and nucleic acid visualization.

Our review of the literature and communication with researchers points to an accelerated use of the TSA method for IHC and *in situ* hybridization detection. In addition, the comparative results of TSA versus *in situ* PCR reported by Luka and Afflerbach[22] make it tempting to foresee widespread use of TSA for low copy nucleic acid detection. Published confirmation and extension of these results over the next few years is likely to define the ultimate usage for this application.

Achieving high sensitivity while retaining low background has made both indirect and direct TSA very valuable techniques. It should be emphasized that these characteristics also offer the user one or more of the following:

- The ability to conserve "hard to obtain" reagents. Typically 10- to >100-fold lower concentrations of primary antibody can be used in IHC or ICC. Similarly, readily synthesized oligonucleotide probes can often be substituted for the multilabeled long probes normally required for low copy number detection using *in situ* hybridization.
- Cost savings. In addition to primary reagents, lower amounts of labeled secondary detectors can be employed.
- Time reduction. The higher sensitivity can be used to reduce the duration of incubation, leading to overall shorter elapsed time in both immunological and probe assays

ACKNOWLEDGMENTS. The authors thank Karl Adler for major contributions to the technology in general and for the CMV IHC and HPV ISH data shown here in particular. Tom Erickson's aid in preparing the figures is greatly appreciated.

REFERENCES

1. Bobrow, M. N., Harris, T. D., Shaughnessy, K. J., and Litt, G. J., 1989, Catalyzed reporter deposition, a novel method of signal amplification. Application to immunoassays, *J. Immunol. Methods* **125:**279.
2. Bobrow, M. N., Shaughnessy, K. J., and Litt, G. J., 1991, Catalyzed reporter deposition, a novel method of signal amplification. Application to membrane immunoassays, *J. Immunol. Methods* **137:**103.
3. Ngo T. T., 1991, Immunoassay, *Curr. Opinions Biotechnol.* **2:**102.
4. Kricka, L. J., 1993, Ultrasensitive immunoassay techniques, *Clin. Biochem.* **26:**325.
5. Diamandis, E. F. and Christopoulos, T. K., 1991, The biotin-(strept)avidin system: Principles and applications in biotechnology, *Clin. Chem.* **37:**625.
6. van Gijlswijk, R. P. M., van Gijlswijk-Janssen, D. J., Raap, A. K., Daha, M. R. and Tanke, H. J., 1996, Enzyme-labelled antibody-avidin conjugates: New flexible and sensitive immunochemical reagents, *J. Immunol. Methods,* **189:**117.
7. Grumbach, I. J. and Veh, R. W., 1995, The SA/rABC technique: A new ABC procedure for detection of antigens at increased sensitivity, *J. Histochem. Cytochem.* **43:**31.
8. Saiki, R. K., Scharf, S., Faloona, F., Mullis, K. B., Horn, G. T., Erlich, H. A. and Arnheim, N., 1985, Enzymatic amplification of β-globin genomic sequences and restriction site analysis of sickle cell anemia, *Science* **230:**1350.

9. Maia, M., Takahashi H., Adler, K., Garlick, R. K. and Wands, J. R., 1995, Development of a two site immuno-PCR assay for hepatitis B surface antigen, *J. Virol. Methods* **52:**273.
10. Larison, K. D., BreMiller, R., Wells, K. S., Clements, I. and Haugland, R. P., 1995, Use of a new fluorogenic phosphatase substrate in immunohistochemical applications, *J. Histochem. Cytochem.* **43:**77.
11. Gross, A. J. and Sizer, I. W., 1959, The oxidation of tyramine, tyrosine, and related compounds by peroxidase, *J. Biol. Chem.* 234:1611.
12. Guilbault, G. G., Brignac, P. J., and Zimmer, M., 1968, Homovanillic acid as a fluorometric substrate for oxidative enzymes, *Anal. Chem.* **40:**190.
13. Guilbault, G. G., Brignac, P. J., and Juneau, M., 1968, New substrates for the fluorometric determination of oxidative enzymes, *Anal. Chem.* **40:**1256.
14. Zaitsu, K. and Ohkura, Y., 1980, New fluorogenic substrates for horseradish peroxidase; Rapid and sensitive assays for hydrogen peroxide and the peroxidase, *Anal. Biochem.* **109:**109.
15. Adams, J. C., 1992, Biotin amplification of biotin and horseradish peroxidase signals in histochemical stains, *J. Histochem. Cytochem.* **40:**1457.
16. van Gijlswijk, R. P. M., Zijlmans, H. J. M. A. A., Wiegant, J., Bobrow, M. N., Erickson, T. J., Adler, K. E., Tanke, H. J., and Raap, A. K., (submitted 1996), Fluorochrome-labelled tyramides: Use in immuno-cytochemistry and fluorescent *in situ* hybridization, *J. Histochem. Cytochem.*
17. ACTG Virology Manual for HIV Laboratories. U.S. Department of Health and Human Services, Public Health Service, NIH Publication No. 94 3828, Compiled by the AIDS Clinical Trials Group Virology Technical Advisory Committee and the Division of AIDS, NIAID, National Institutes of Health, September 1994.
18. Sissors, D., 1996, Unpublished report, NEN Life Science Products, Boston, MA.
19. Lyamuya, E., Bredberg-Raden, U., Massawe, A., Urassa, E., Kawo, G., Msemo, G., Kazimoto, T., Ostborn, A., Karlsson, K., Mhalu, F., and Biberfeld, G., 1996, Performance of a modified HIV-1 p24 antigen assay for early diagnosis of HIV-1 infection in infants and prediction of mother-to-infant transmission of HIV-1 in Dar es Salaam, Tanzania, *J. Acquired Immune Defic. Syndr. Hum. Retrovirol.* **12:**421.
20. Schupbach, J., Flepp, M., Pontelli, D., Tomasik, Z., Luthy, R., and Boni, J., 1996, Heat-mediated immune complex dissociation and enzyme immunoassay signal amplification render antigen p24 detection in plasma as sensitive as HIV-1 RNA detection by polymerase chain reaction, *AIDS* **10:**1085.
21. Deelder, A. M., Qian, Z. L., Kremsner, P. G., Acosta, L., Rabello, A. L. T., Enyong, P., Simarro, P. P., van Etten, E. C. M., Krijger, F. W., Rotmans, J. P., Fillie, Y. E., de Jonge, N., Agnew, A. M., and van Lieshout, L., 1994, Quantitative diagnosis of Schistosoma infections by measurement of circulating antigens in serum and urine, *Trop. Geogr. Med.* **46:**233.
22. Luka, J. and Afflerbach, C., 1997, Detection of low copy number viral genomes in lymphoctyes and tissues using tyramide signal amplification. *BioTechniques*, in press.
23. Zecchini, V., Cacoethes, D., Gretry, J., Lemarite, M., and Dommes, J., 1995. Detection of nucleic acids in the attomole range using polybiotinylated oligonucleotide probes. *BioTechniques* **19:**286.

11

Ribotyping as a Tool for Molecular Epidemiology

SUSAN B. HUNTER and BALA SWAMINATHAN

1. INTRODUCTION

DNA restriction fragment length polymorphism (RFLP)-based methods have become invaluable laboratory tools for identifying and typing pathogenic bacteria to support epidemiological investigations of infectious diseases. If these investigations include a molecular typing component, this approach is often called "molecular epidemiology." In the 1980s plasmid profiles and plasmid RFLPs were used extensively for molecular epidemiology. The disadvantages of plasmid-based typing were that not all pathogenic bacteria harbor plasmids, that there may be insufficient variability in the plasmid(s) carried by some pathogens, and that plasmids may not be stable markers in some other pathogens. As simple methods to characterize genomic DNA were developed, they were increasingly adopted for molecular epidemiology, often supplementing or replacing plasmid-based typing methods. Genomic DNA RFLP methods are universally applicable, always have high discriminating ability, and are stable. However, restriction of genomic DNA of bacteria by using frequently cutting restriction enzymes often yielded more than 100 fragments that generated relatively complex patterns after electrophore-

Use of trade names is for identification only and does not imply endorsement by the Public Health Service or by the U.S. Department of Health and Human Services.

SUSAN B. HUNTER and BALA SWAMINATHAN • Foodborne and Diarrheal Diseases Branch, Division of Bacterial and Mycotic Diseases, National Center for Infectious Diseases, Centers for Disease Control and Prevention, Atlanta, Georgia 30333.

Rapid Detection of Infectious Agents, edited by Specter *et al.* Plenum Press, New York, 1998.

tic separation. Therefore, attempts were made to simplify the patterns obtained by genomic DNA restriction analysis to facilitate comparing large numbers of isolates by visual or computer-assisted methods. One such approach involves the Southern hybridization of the genomic DNA restriction fragments with a probe that has multiple homologous sequences on the chromosome. Examples of probes used for this approach include insertion elements, virulence-associated genes, and ribosomal RNA.

2. RIBOTYPING—THE METHOD

Ribotyping is a term coined by Stull *et al.*[1] to describe a Southern hybridization procedure in which genomic DNA restriction fragments are hybridized with

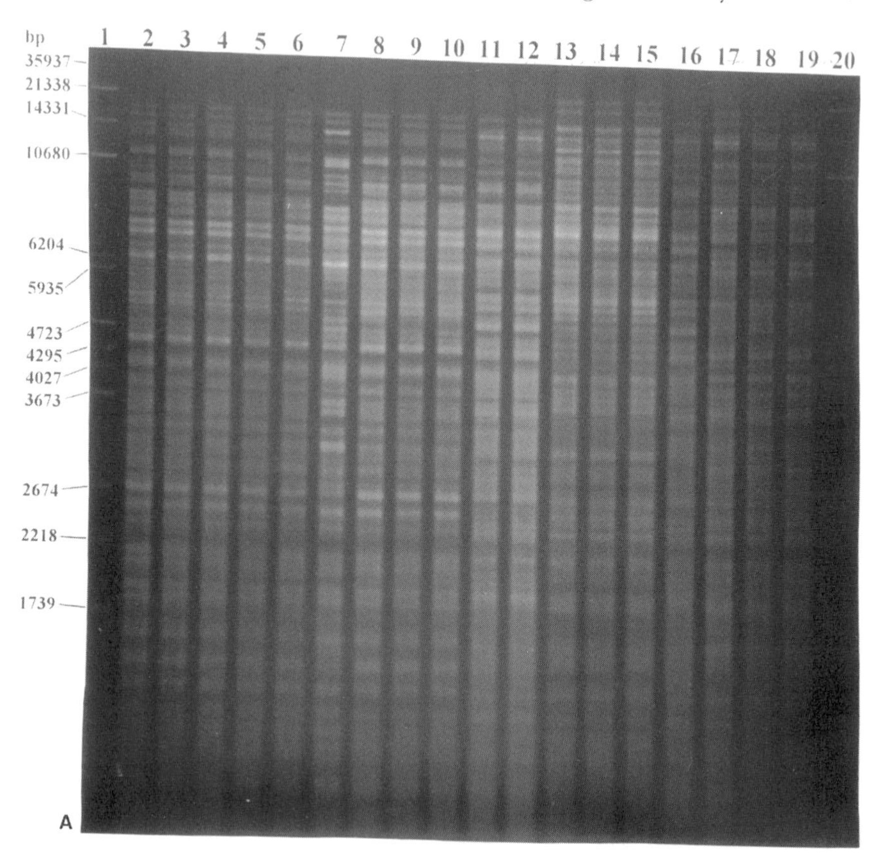

FIGURE 1. DNA restriction profiles of (A) 18 isolates of *Corynebacterium diphtheriae* and (B) their corresponding ribotype patterns.

a probe derived from 16S and 23S ribosomal RNA. After the hybridization step, only those genomic DNA fragments that contain some or all ribosomal DNA are rendered visible (Fig. 1). Thus, a DNA-RFLP pattern that contains 100 or more DNA fragments after restriction and electrophoresis can be simplified to about 10–20 fragments of different sizes after hybridization with rRNA or rDNA. Because the ribosomes are extensively involved in protein synthesis and assembly, most bacteria contain multiple ribosomal operons scattered throughout the chromosome and thus generate an adequate number of fragments suitable for strain identification and discrimination.[2–5] Figure 2 is a representation of the genetic map of *Escherichia coli* with the locations of the ribosomal operons indicated. Some slowly growing bacteria (e.g., *Mycobacterium* spp.) may contain only one or two ribosomal operons. Ribotyping is not useful for application to these bacteria. A major advantage of ribotyping is that a single rRNA/rDNA probe can be used to type a wide variety of bacteria because ribosomal RNA is made up of highly conserved and highly variable regions. The highly conserved regions are needed to maintain the secondary structure of the rRNA, which is required for the constitution of ribosomes. The highly conserved regions also enable using a single rRNA probe as a universal probe. *E. coli* rRNA is used extensively for ribotyping gram-positive and gram-negative bacteria.

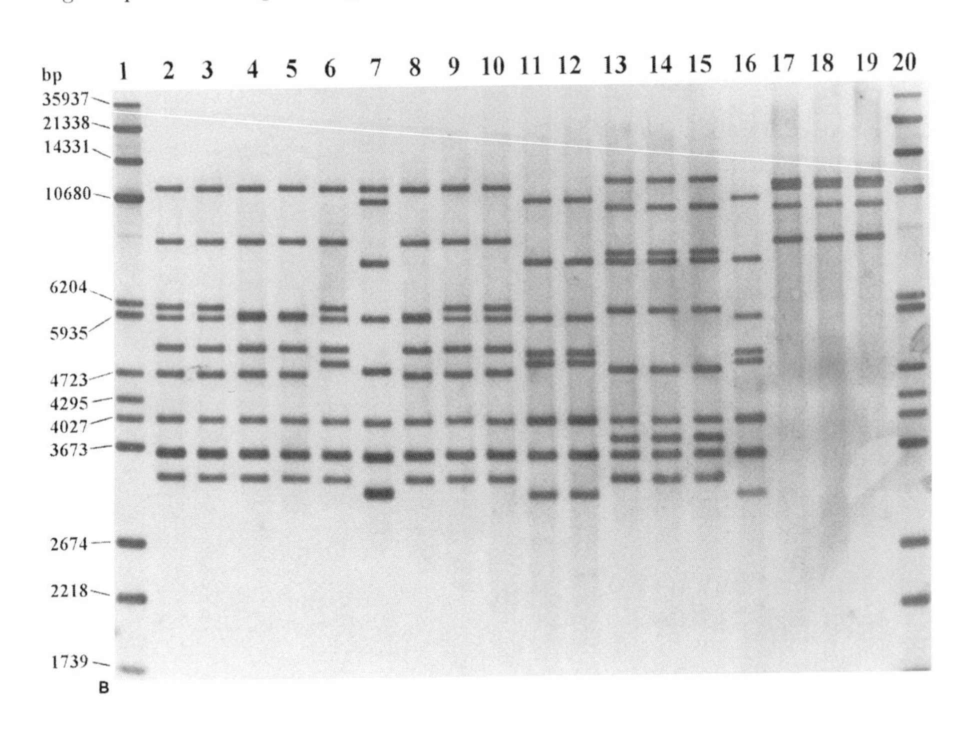

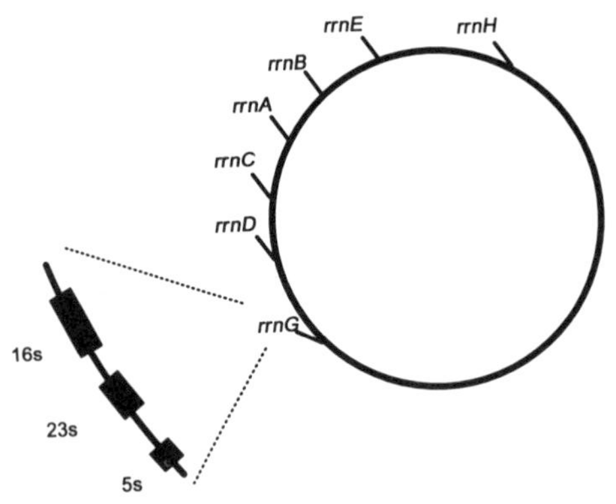

FIGURE 2. Physical map of *Escherichia coli* genome and the locations of the ribosomal operons.

3. PROBES AND LABELS USED FOR RIBOTYPING

Initial experiments used 16S and 23S rRNA of *E. coli.*[1,2] However, commercially available 16S and 23S rRNA of *E. coli* is contaminated with 5S rRNA.[6] RNA is also easily degraded by ribonucleases that are present everywhere if stringent precautions are not taken. Therefore, recent ribotyping applications have used ribosomal DNA. Plasmid pKK3535, constructed by cloning the *rrnB* operon of *E. coli* into plasmid vector pBR322, has been used by several investigators as the source of a probe for ribotyping.[5,7] The plasmid pKK3535 is restricted with an appropriate enzyme to remove the rDNA from the vector sequence before labeling it for ribotyping. Alternately, cDNA may be obtained by reverse transcription of 16 + 23S rRNA.[8]

In the early ribotyping efforts in the 1980s, the probe was usually labeled with radioisotopes (^{32}P).[1,2] Later, the radioactive label was almost entirely replaced by a variety of nonisotopic labels, including biotin,[7] digoxigenin,[8–10] and horseradish peroxidase-polyethyleneimine complex.[11,12] An acetylaminofluorene-labeled 16 + 23S rRNA kit is commercially available from Eurogentec S.A., Belgium.[5] The nonisotopic labels are detected by using appropriate chromogenic or chemiluminescent detection reagents. An advantage of chemiluminescent detection is that the membrane with the immobilized DNA fragments can be reprobed after the previous probe is stripped by physical or chemical methods.

4. RIBOTYPING APPLICATIONS IN MOLECULAR EPIDEMIOLOGY

Ribotyping has been applied to the epidemiologic typing of a large number of gram-positive and gram-negative bacteria and chlamydia. The use of ribotyping for epidemiological investigations of nosocomial outbreaks has been recently reviewed.[6]

Among gram-positive bacteria, ribotyping has been used to subtype *Staphylococcus aureus, S. epidermidis,* other coagulase-negative staphylococci, group B *Streptococcus* (GBS), and *Listeria monocytogenes.* In comparing restriction endonuclease analysis (REA) and ribotyping for typing 76 GBS isolates associated with human invasive disease, Blumberg *et al.* found that the isolates were divided into nine ribotypes. Serotypes Ia/c, II/c and III are distinguishable by their ribotype patterns.[13] Although REA and ribotyping shows indistinguishable patterns for epidemiologically related isolates, REA is more discriminatory for epidemiologically unrelated isolates.

Blumberg *et al.*[14] applied ribotyping to study the emergence and spread of ciprofloxacin-resistant *S. aureus.* Many of these isolates could not be characterized by phage typing. Using ribotyping, they showed that several strains of methicillin-resistant *S. aureus* (MRSA) exist in their medical institution and that ciprofloxacin resistance emerged in multiple strains of MRSA, in contrast to the situation in Tel Aviv Medical Center, where a single MRSA strain was identified by ribotyping as responsible for the ciprofloxacin resistance. Thus, ribotyping offered a means to type non-phage-typable MRSA and methicillin-sensitive *S. aureus.*[15] Ribotyping is of value for typing coagulase-negative staphylococci[16,17] although REA is more discriminating than ribotyping. Shayegani *et al.*[18] used ribotyping to characterize *S. epidermidis* isolates from a patient who developed sepsis after platelet transfusion from all eight units of platelets she received.

Ribotyping was applied to study the epidemiology of outbreak-associated and sporadic human listeriosis and to delineate the role of foods in transmitting listeriosis.[19] However, other molecular methods have greater discriminating power than ribotyping for *L. monocytogenes,* especially isolates belonging to serotype 4b.[10]

For gram-negative bacteria, ribotyping was successfully applied to subtype 133 *E. coli* isolates from various extraintestinal infections or stools from healthy persons. These isolates represented 9 O groups and 34 O:K:H serotypes. They were divided into 20 ribotypes. The ribotypes generally correlated with the O:K:H serotypes. Ribotype variations within a serotype were seen primarily in strains with K5 capsule.[20] Ribotyping is also useful for subtyping *Shigella sonnei*[4] and *S. flexneri.*[9]

Koblavi *et al.*[21] ribotyped 89 *Vibrio cholerae* strains from different geographic locations. The 89 strains were divided into 17 ribotypes. There was no correlation between serotype and ribotype. However, the classical and El Tor biotypes had unique ribotype patterns associated with each, and some ribotypes were found only in certain geographic locations. An extensive evaluation of ribotyping for subtyping *V. cholerae* O1 was carried out by Popovic *et al.*[22] They ribotyped 214 clinical and environmental isolates collected from 35 countries and 14 U.S. states, over a 60-year period. Seven different but highly similar ribotypes were found among 16 strains of the classical biotype, whereas 20 ribotypes were identified among nearly 200 strains of the El Tor biotype. Some ribotypes were associated with isolates from specific geographic locations. Because the ribotypes were stable and reproducible, Popovic *et al.* suggested that they could assist in tracing the movement of *V. cholerae* O1 strains and identifying their geographic origins.

V. cholerae O139, a new serogroup of *V. cholerae,* caused epidemic cholera in Bangladesh and India in 1992–93. Faruque *et al.*[23] used ribotyping to show that all 29 isolates from Bangladesh and India were a single ribotype and that they were closely related to the El Tor biotype of *V. cholerae* O1.

Guiyoule *et al.*[3] ribotyped 70 strains of *Yersinia pestis* isolated on five continents over a 72-year period. The strains were divided into 16 ribotypes, but 66% of the strains were clustered in two ribotypes. Greater genetic heterogeneity was found among African isolates when compared with Asian isolates. North and South American isolates were of a single ribotype. On the basis of their work and those of others, Guiyoule *et al.* suggested that one main clone of *Y. pestis* spread from Central Asia to Central Africa and caused Justinian's plague. Further, they hypothesized that a variant of the same clone spread from Central Asia to the Crimea and was responsible for the Black Death. These inferences could be made because ribotypes were highly stable and reproducible.

Yersinia enterocolitica is a major enteric pathogen associated with a wide variety of acute and chronic clinical syndromes. *Y. enterocolitica* infections are most commonly manifested as enterocolitis in young children and older adults. A pseudoappendicular syndrome primarily occurs in older children and young adults. In addition, *Y. enterocolitica* infections are associated with a variety of chronic arthritic and rheumatoid syndromes. This organism has also been implicated in sepsis associated with transfusion of erythrocytes in the United States and Europe. Blumberg *et al.*[24] used ribotyping to show that a limited number of *Y. enterocolitica* strains of serotype O:3 have been disseminated within the United States and globally. They also demonstrated the utility of ribotyping for epidemiologically investigating an outbreak caused by consumption of contaminated hog intestines (chitterlings).

5. USE OF MULTIPLE RESTRICTION ENZYMES

Multiple enzymes are often required to achieve adequate discrimination between epidemiologically unrelated isolates of bacteria. Yogev *et al.*[25] showed that two enzymes (*Eco*RI and *Pst*I) were necessary to obtain epidemiologically relevant discrimination between *Acholeplasma laidlawii* isolates. Other investigators have used two enzymes to ribotype MRSA,[15] epidemic-associated *Campylobacter* *jejuni*/*coli* strains,[26] and clinical isolates of *Haemophilus ducreyi*.[27] Woods *et al.*[28] used three enzymes to type *N. meningitidis* serogroup C isolates. The 44 isolates were divided into 7 ribotypes using *Cla*I, 8 ribotypes using *Xho*I, and nine ribotypes using *Eco*RI. When the data for the three enzymes were combined, 17 different ribotypes were obtained. Bangsborg *et al.*[29] found that the discriminating power of ribotyping increases by combining the results obtained with four restriction enzymes. Rozee *et al.*[30] also demonstrated increased ability to discriminate between *Pseudomonas cepacia* isolates by using five restriction enzymes. Interestingly, the *Eco*RI ribotypes are stable for sequential isolates from the same patient, whereas ribotypes obtained by using other enzymes vary, particularly during the acute infection phase. Mendoza *et al.*[9] used six different restriction enzymes to type sporadic and outbreak-associated isolates of *S. sonnei* and *S. flexneri*. The discrimination index (DI) obtained with a single enzyme ranged from 0.14 to 0.54 for *S. sonnei* and 0.63 to 0.71 for *S. flexneri*. Combining the results of two or more enzymes increases the DI. A DI of 0.91 for *S. sonnei* is achievable by combining the ribotyping results obtained by using six enzymes. A DI of 0.81 is obtained for *S. flexneri* by using certain combinations of three or more enzymes. While the DI increases to acceptable levels (DI > 0.90) by using multiple enzymes for ribotyping, this results in a significantly increased workload for the laboratory. Also, the use of multiple enzymes does not significantly increase the DI for some species of bacteria.[19]

Hinojosa-Ahumada *et al.*[4] devised a two-step procedure for ribotyping *S. sonnei* with multiple enzymes. They used a single enzyme (*Sal*I) to type all isolates. Depending on the pattern obtained with *Sal*I, a suitable second enzyme (*Pvu*II, *Sma*I, or *Sst*I) was used to discriminate further among the isolates. This approach increases the DI to an acceptable level while keeping the laboratory workload under control.

6. COMPARISON OF RIBOTYPING WITH OTHER MOLECULAR SUBTYPING METHODS

A qualitative comparison of the discriminating power of ribotyping with that of other molecular subtyping methods for a representative group of bacteria is presented in Table I. Ribotyping has equal or better discriminating ability compared with some phenotypic and molecular subtyping methods for some pathogenic bacteria, whereas it is clearly inferior to other subtyping methods, such as DNA macrorestriction analysis by pulsed-field gel electrophoresis (PFGE) and random amplified polymorphic DNA analysis. Ribotyping is useful for those species for which significant numbers of isolates are untypeable by PFGE.

TABLE I
Comparative Evaluation of Ribotyping and Other Subtyping Methods

Species	Restriction enzyme(s) used for ribotyping	Method(s) compared[a]	Discriminating ability	Comments	Reference
Acinetobacter calcoaceticus-baumanii group	*Eco*RI, *Cla*I	PFGE	PFGE>RT	73/73 strains typeable by RT; 70 by PFGE	42
Campylobacter jejuni	*Hae*III, *Pst*I, *Kpn*I	Bacteriophage typing, PFGE	PFGE>phage typing>RT	22% of isolates untypeable by PFGE	43
Campylobacter jejuni	*Pvu*II, *Pst*I	Plasmid profiles, MEE, REA	MEE=REA=RT >plasmid profiles		26
Clostridium difficile	*Hind*III	PFGE,RAPD	PFGE>RAPD>RT		44
Clostridium difficile	*Hind*III	REA,PFGE	PFGE=REA>RT	Some isolates untypeable by PFGE	45
Corynebacterium diptheriae	*Bst*EII	PFGE	RT>PFGE	PFGE less reproducible than RT	46
Enterobacter aerogenes	*Eco*RI, *Pvu*II	AP-PCR	AP-PCR=RT		47

Enterococci	?	Biochemical fingerprinting and PFGE	PFGE>BCHM>RT		48
Legionella pneumophila	*Eco*RI, *Pst*I	AFLP	AFLP=RT		49
Listeria monocytogenes	*Eco*RI	Phage typing, MEE, REA, PFGE, RAPD	PFGE=RAPD> REA>MEE>RT	RT has low discrimination for serotype 4b	10
Neisseria meningitidis serogroup B	*Cla*I	McAb serotyping, MEE, ITS-PCR/RFLP, PFGE	PFGE>MEE>RT > McAb serotyping >ITS-PCR/RFLP		36
Salmonella serotype Enteritidis	*Sma*I, *Sph*I	PFGE	Both have low discrimination		50
Salmonella serotype Typhi	*Pst*I	Phage typing	RT>phage typing		7
Salmonella serotype Typhimurium	*Bgl*II, *Hind*III, *Pvu*II, *Sma*I	IS200 typing	IS200>RT		51
Serratia marcescens	*Eco*RI, *Hind*III	Biotyping	RT>biotyping	Ribotype patterns shared between biotypes	52

[a]PFGE: pulsed-field gel electrophoresis; RT: ribotyping; MEE: multilocus enzyme electrophoresis; REA: restriction endonuclease analysis; RAPD: random amplified polymorphic DNA; AP-PCR: arbitrarily primed polymerase chain reaction; AFLP: amplified fragment length polymorphism; McAb: monoclonal antibody; ITS-PCR/RFLP: internally transcribed spacer polymerase chain reaction/restriction fragment polymorphism; IS 200: insertion sequence 200.

7. POLYMERASE CHAIN REACTION RIBOTYPING

Polymerase chain reaction (PCR) ribotyping is an attempt to simplify ribotyping. In this approach, polymorphisms within the ribosomal operon are used to type strains of bacteria. Kostman *et al.*[31] amplified the intergenic region between the genes encoding 16S and 23S rRNA and demonstrated that the polymorphisms observed in this fragment could be used to type *Pseudomonas cepacia*. Later, they reported that the procedure is broadly applicable by testing the method with *Staphylococcus aureus, Enterococcus fecium, Enterococcus fecalis, Enterobacter aerogenes, Enterobacter cloacae,* and urinary isolates of *E. coli.*[32] PCR ribotyping takes advantage of the variation in the length and sequence of the intergenic spacer region within different ribosomal operons of an isolate. Cross-hybridization of different PCR amplicons with homologous termini that contain variable internal sequences also results in the formation of heteroduplex DNA that migrates more slowly than the corresponding homologous sequences.[32] Nastasi and Mammina[33] applied PCR ribotyping to 45 isolates of *Salmonella* serotype Typhimurium and compared the results with those of conventional ribotyping. They found 10 different patterns by PCR ribotyping and 9 patterns with conventional ribotyping. Epidemiologically related and unrelated isolates were correctly identified by both methods. PCR ribotyping results agreed completely with those of arbitrarily primed PCR for 43 isolates of *Clostridium difficile* from several Polish hospitals.[34] Smith-Vaughan *et al.*[35] amplified the entire 5.5 kbp of the ribosomal operon of the genus *Haemophilus* and determined the RFLP of these amplicons. They applied this method to characterize 49 nonserotypeable nasopharyngeal *H. influenzae* isolates that had been previously typed by conventional ribotyping. Once again, they found remarkably similar discrimination of the isolates by both methods. Swaminathan *et al.*[36] found that using a PCR-restriction fragment length polymorphism method based on amplifying the internally transcribed spacer region of *Neisseria meningitidis* serogroup B followed by restriction of the amplified fragment was less discriminating than other molecular subtyping methods. Although PCR ribotyping is a rapid method, its utility for each species needs to be validated.

8. AUTOMATION OF RIBOTYPING

Ribotyping is a complex and tedious laboratory procedure that involves multiple steps of DNA isolation and purification, restriction of DNA, electrophoretic separation of DNA restriction fragments, transfer of DNA fragments to a membrane matrix and immobilization, purification of the probe and labeling with a radioisotope or a nonradioactive reporter molecule, hybridization of the labeled probe with the DNA restriction fragments, visualization of the hybridized fragments, and finally analyzing and interpreting the patterns. The entire procedure has been automated in the RiboPrinter™ Microbial Characterization System (Qualicon, L.L.C., Wilmington, Delaware).[37] Up to eight cultures can be processed in one batch. Although the RiboPrinter™ requires eight hours to complete the procedure, up to three additional batches may be loaded into the machine at intervals of two hours. The first version of the RiboPrinter™ performed only typing using *Eco*RI enzyme, which severely limited the instrument's capability. An updated version of the software allows the use of other restriction enzymes. However, the operator must manually purify and restrict the DNA and load the restriction digest in the RiboPrinter™. This protocol is amenable for use with bacteria that are not efficiently lysed by the automated lysis protocol of the RiboPrinter, which has been optimized for *Listeria.*

The RiboPrinter™ is particularly useful for ribotyping large numbers of isolates. Ryser *et al.*[38] used the RiboPrinter™ to ribotype 608 isolates of *L. monocytogenes* in a study designed to compare the ability of primary enrichment media to recover different genotypes of *L. monocytogenes* from raw meats and poultry. The RiboPrinter™ should also be useful for screening a large panel of restriction enzymes to assess their usefulness in ribotyping a specific bacterial species.

9. OTHER APPLICATIONS OF RIBOTYPING

Ribotyping has been applied to identify bacteria at the species level and for taxonomic studies of bacteria.[2,39,40] Hubner *et al.*[41] used ribotyping data obtained from more than 1300 strains of *L. monocytogenes* to predict potential *Eco*RI cleavage sites in the ribosomal operons of this organism and to predict theoretically possible, but as yet undiscovered, ribotypes of *L. monocytogenes.*

10. SUMMARY

Ribotyping was a powerful tool for molecular epidemiology in the 1980s and early 1990s but has been largely superseded by PFGE analysis and PCR-based typing methods that have higher discriminating ability. Nevertheless, ribotyping is a useful method for characterizing bacteria and will continue to be applied to identifying and classifying bacteria. PCR ribotyping methods offer a rapid and simple means of subtyping certain bacteria for which their discriminating ability is acceptable. In large reference and quality control laboratories, an automated ribotyping instrument may facilitate the use of ribotyping for identification and classification of bacteria and for preliminary subtyping in epidemiological investigations before the application of time-consuming and labor-intensive methods, such as PFGE.

ACKNOWLEDGMENTS. We thank Lewis M. Graves for providing Fig. 1(A) and Fig. 1(B). We thank Thomas Donkar and Eric Renner for their assistance with the literature review for this chapter.

REFERENCES

1. Stull, T. L., LiPuma, J. J., and Edlind, T. D., 1988, A broad-spectrum probe for molecular epidemiology of bacteria: Ribosomal RNA, *J. Infect. Dis.* **157:**280–286.
2. Grimont, F. and Grimont, P. A. D., 1986, Ribosomal ribonucleic acid gene restriction patterns as potential taxonomic tools, *Ann. Inst. Pasteur/Microbiol.* **137 B:**165–175.
3. Guiyoule, A., Grimont, F., Iteman, I., Grimont, P. A. D., Lefevre, M., and Carniel, E., 1994, Plague pandemics investigated by ribotyping of *Yersinia pestis* strains, *J. Clin. Microbiol.* **32:**634–641.
4. Hinojosa-Ahumada, M., Swaminathan, B., Hunter, S. B., Cameron, D. N., Kiehlbauch, J. A., Wachsmuth, I. K., and Strockbrine, N. A., 1991, Restriction fragment length polymorphisms in rRNA operons for subtyping *Shigella sonnei*, *J. Clin. Microbiol.* **29:**2380–2384.
5. Swaminathan, B. and Matar, G. M., 1993, Molecular typing methods, in: *Diagnostic Molecular Microbiology: Principles and Applications* (D. H. Persing, T. F. Smith, F. C. Tenover, and T. J. White, eds.), American Society for Microbiology, Washington, D.C. pp. 26–50.
6. Bingen, E. H., Denamur, E., and Elion, J., 1994, Use of ribotyping in epidemiological surveillance of nosocomial outbreaks, *Clin. Microbiol. Rev.* **7:**311–327.
7. Altwegg, M., Hickman-Brenner, F. W., and Farmer J. J. III, 1989, Ribosomal RNA gene restriction patterns provide increased sensitivity for typing *Salmonella typhi* strains, *J. Infect. Dis.* **160:**145–149.
8. Popovic, T., Bopp, C. A., Olsvik, O., and Keihlbauch, J., 1993, Ribotyping in molecular epidemiology, in: *Diagnostic Molecular Microbiology: Principles and Applications* (D. H. Persing, T. F. Smith, F. C. Tenover, and T. J. White, eds.), American Society for Microbiology, Washington, D.C. pp. 573–583.
9. Mendoza, M. C., Martin, M. C., and Gonzalez-Hevia, M. A., 1996, Usefulness of ribotyping in a molecular epidemiology study of shigellosis, *Epidemiol. Infect.* **116:**127–135.
10. Swaminathan, B., Hunter, S. B., Desmarchelier, P. M., Gerner-Smidt, P., Graves, L. M., Harlander, S., Hubner, R., Jacquet, C., Pederson, B., Reineccius, K., Ridley, A., Saunders, N. A., and Webster, J. A., 1996, WHO-sponsored international collaborative study to evaluate methods for subtyping *Listeria monocytogenes:* Restriction fragment length polymorphism (RFLP) analysis using ribotyping and Southern hybridization with two probes derived from *L. monocytogenes* chromosome, *Int. J. Food Microbiol.* **32:**263–278.
11. Gustaferro, C. A. and Persing, D. H., 1992, Chemiluminescent universal probe for bacteria ribotyping, *J. Clin. Microbiol.* **30:**1039–1041.
12. Gustaferro, C. A., 1993, Chemiluminescent ribotyping, in: *Diagnostic Molecular Microbiology: Principles and Applications* (D. H. Persing, T. F. Smith, F. C. Tenover, and T. J. White, eds.), American Society for Microbiology, Washington, D.C. pp. 584–589.
13. Blumberg, H. M., Stephens, D. S., Licitra, C., Pigott, N., Facklam, R., Swaminathan, B., and Wachsmuth, I. K., 1992, Molecular epidemiology of Group B streptococcal infections: Use of restriction endonuclease analysis of chromosomal DNA and DNA restriction fragment length polymorphisms of ribosomal RNA genes (ribotyping), *J. Infect. Dis.* **166:**574–579.
14. Blumberg, H. M., Rimland, D., Kiehlbauch, J. A., Terry, P. M., and Wachsmuth, I. K., 1992, Epidemiologic typing of *Staphyloccous aureus* by DNA restriction fragment length polymorphisms of

rRNA genes: Elucidation of the clonal nature of a group of bacteriophage-nontypeable, ciprofloxacin-resistant, methicillin-susceptible *S. aureus isolates, J. Clin. Microbiol.* **30:**362–369.

15. Preheim, L., Pitcher, D., Owen, R., and Cookson, B., 1991, Typing of methicillin-resistant and -susceptible *Staphylococcus aureus* strains by ribosomal RNA gene restriction patterns using a biotinylated probe, *Eur. J. Clin. Microbiol. Infect. Dis.* **5:**428–436.
16. Hawcroft, D. M. and Geary, C., 1996, The use of a nonradioactively labeled probe system in an electrophoretic ribotyping method for the differentiation of strains of coagulase-negative staphylococci, *Electrophoresis* **17:**55–57.
17. Bialkowska-Hobrzanska, H., Harry, V., Jaskot, D., and Hammerberg, O., 1990, Tying of coagulase-negative staphylococci by Southern hybridization of chromosomal DNA fingerprints using a ribosomal RNA probe, *Eur. J. Clin. Microbiol. Infect. Dis.* **9:**588–594.
18. Shayegani, M., Parsons, L. M., Waring, A., Donhowe, J., Goering, R., Archinal, W. A., and Linden, J., 1991, Molecular relatedness of *Staphylococcus epidermidis* isolates obtained during a platelet transfusion-associated episode of sepsis, *J. Clin. Microbiol.* **29:**2768–2773.
19. Baloga, A. O. and Harlander, S. K., 1991, Comparison of methods for discrimination between strains of *Listeria monocytogenes* from epidemiological surveys, *Appl. Environ. Microbiol.* **57:**2324–2331.
20. Tarkka, E., Ahman, H., and Siitonen, A., 1994, Ribotyping as an epidemiologic tool for *Escherichia coli, Epidemiol. Infect.* **112:**263–274.
21. Koblavi, S., Grimont, F., and Grimont, P. A. D., 1990, Clonal diversity of *Vibrio cholerae* O1 evidenced by rRNA gene restriction patterns, *Res. Microbiol.* **141:**645–657.
22. Popovic, T., Bopp, C., Olsvik, O., and Wachsmuth, K., 1993, Epidemiology application of a standardized ribotype scheme for *Vibrio cholerae* O1, *J. Clin. Microbiol.* **31:**2472–2482.
23. Faruque, S. M., Abdul Alim, A. R. M., Roy, S. K., Khan, F., Nair, G. B., Sack, R. B., and Albert, M. J., 1994, Molecular analysis of rRNA and cholera toxin genes carried by the new epidemic strain of toxigenic *Vibrio cholerae* O139 synonym Bengal, *J. Clin. Microbiol.* **32:**1050–1053.
24. Blumberg, H. M., Kiehlbauch, J. A., and Wachsmuth, I. K., 1991, Molecular epidemiology of *Yersinia enterocolitica* O:3 infections: Use of chromosomal DNA restriction fragment length polymorphisms of rRNA genes, *J. Clin. Microbiol.* **29:**2368–2374.
25. Yogev, D., Levisohn, S., Kleven, S. H., Halachimi, D., and Razin, S., 1988, Ribosomal RNA gene probes to detect intraspecies heterogeneity in *Mycoplasma gallisepticum* and *M. synoviae, Avian Dis.* **32:**220–231.
26. Patton, C. M., Wachsmuth, I. K., Evins, G. M., Kiehlbauch, J. A., Plikaytis, B. D., Troup, N., Tompkins, L., and Lior, H., 1991, Evaluation of 10 methods to distinguish epidemic-associated *Campylobacter* strains, *J. Clin. Microbiol.* **29:**680–688.
27. Sarafian, S. K., Woods, T. C., Knapp, J. S., Swaminathan, B., and Morse, S. A., 1991, Molecular characterization of *Haemophilus ducreyi* by ribosomal DNA fingerprinting, *J. Clin. Microbiol.* **29:**1948–1954.
28. Woods, T. C., Helsel, L. O., Swaminathan, B., Bibb, W. F., Pinner, R. W., Gellin, B. G., Collin, S. F., Waterman, S. H., Reeves, M. W., Brenner, D. J., and Broome, C. V., 1992, Characterization of *Neisseria meningitidis* serogroup C by multilocus enzyme electrophoresis and ribosomal DNA restriction profiles (ribotyping), *J. Clin. Microbiol.* **30:**132–137.
29. Bangsborg, J. M., Gerner-Smidt, P., Colding, H., Fiehn, N.-E., Bruun, B., and Hoiby, N., 1995, Restriction fragment length polymorphism of rRNA genes for molecular typing of members of the family *Legionellaceae, J. Clin. Microbiol.* **33:**402–406.
30. Rozee, K. R., Haase, D., Macdonald, N. E., and Johnson, W., 1994, Comparison by extended ribotyping of *Pseudomonas cepacia* isolated from cystic fibrosis patients with acute and chronic infections, *Diagn. Microbiol. Infect. Dis.* **20:**181–186.
31. Kostman, J. R., Edlind, T. D., LiPuma, J. J., and Stull, T. L., 1992, Molecular epidemiology of

Pseudomonas cepacia determined by polymerase chain reaction ribotyping, *J. Clin. Microbiol.* **30:**2084–2087.

32. Kostman, J. R., Alden, M. B., Mair, M., Edlind, T. D., LiPuma, J. L., and Stull, T. L., 1995, A universal approach to bacterial molecular epidemiology by polymerase chain reaction ribotyping, *J. Infect. Dis.* **171:**204–208.
33. Nastasi, A. and Mammina, C., 1995, Epidemiological evaluation by PCR ribotyping of sporadic and outbreak-associated strains of *Salmonella enterica* serotype Typhimurium, Res. Microbiol. **146:**99–106.
34. Martirosian, G., Kuipers, S., Verbrugh, H., van Belkum, A., and Meisel-Mikolajczyk, F., 1995, PCR ribotyping and arbitrarily primed PCR for typing strains of *Clostridium difficile* from a Polish maternity hospital, *J. Clin. Microbiol.* **33:**2016–2021.
35. Smith-Vaughan, H. C., Sriprakash, K. S., Matthews, J. D., and Kemp, D. J., 1995, Long PCR-ribotyping of nontypeable *Haemophilus influenzae*, *J. Clin. Microbiol.* **33:**1192–1195.
36. Swaminathan, B., Matar, G. M., Reeves, M. W., Graves, L. M., Ajello, G., Bibb, W. F., Helsel, L. O., Morales, M., Dronavalli, H., El-Swify, M., DeWitt, W., and Hunter, S. B., 1996, Molecular subtyping of *Neisseria meningitidis* serogroup B: Comparison of five methods, *J. Clin. Microbiol.* **34:**1468–1473.
37. Bruce, J., 1996, Automated system rapidly identifies and characterizes microorganisms in food, *Food Technol.* **50** (1): 77–81.
38. Ryser, E. T., Arimi, S. M., Bunduki, M. M. C., and Donnelly, C. W., 1996, Recovery of different *Listeria* ribotypes from naturally contaminated, raw refrigerated meat and poultry products with two primary enrichment media, *Appl. Environ. Microbiol.* **62:**1781–1787.
39. Yogev, D., Halachmi, D., Kenny, G. E., and Razin, S., 1988, Distinction of species and strains of mycoplasmas (mollicutes) by genomic DNA fingerprints with an rRNA gene probe, *J. Clin. Microbiol.* **26:**1198–1201.
40. Laurent, F., Carlotti, A., Boiron, P., Villard, J., and Freney, J., 1996, Ribotyping: A tool for taxonomy and identification of the *Nocardia asteroides* complex species, *J. Clin. Microbiol.* **34:**1079–1082.
41. Hubner, R. J., Cole, E. M., Bruce, J. L., McDowell, C. I., and Webster, J. A., 1995, Types of *Listeria monocytogenes* predicted by the positions of *Eco*RI cleavage sites relative to ribosomal RNA sequences, *Proc. Natl. Acad. Sci. USA* **92:**5234–5238.
42. Seifert, H. and Gerner-Smidt, P., 1995, Comparison of ribotyping and pulsed-field gel electrophoresis for molecular typing of *Acinetobacter* isolates, *J. Clin. Microbiol.* **33:**1402–1407.
43. Gibson, J. R., Fitzgerald, C., and Owen, R. J., 1995, Comparison of PFGE, ribotyping and phage-typing in the epidemiological analysis of *Campylobacter jejuni* serotype HS2 infections, *Epidemiol. Infect.* **115:**215–225.
44. Kristjansson, M., Samore, M. H., Gerding, D. N., DeGirolami, P. C., Bettin, K. M., Karchmer, A. W., and Arbeit, R. D., 1994, Comparison of restriction endonuclease analysis, ribotyping, and pulsed-field gel electrophoresis for molecular differentiation of *Clostridium difficile* strains, *J. Clin. Microbiol.* **32:**1963–1969.
45. Chachaty, E., Saulnier, P., Martin, A., Mario, N., and Andremont, A., 1994, Comparison for ribotyping, pulsed-field gel electrophoresis and random amplified polymorphic DNA for typing *Clostridium difficile* strains, *FEMS Microbiol. Lett.* **122:**61–68.
46. De Zoysa, A., Efstratiou, A. George, R. C., Jahkola, M., Vuopio-Varkila, J., Deshevoli, S., Tseneva, G., and Rikushin, Y., 1995, Molecular epidemiology of *Corynebacterium diphtheriae* from northwestern Russia and surrounding countries studied by using ribotyping and pulsed-field gel electrophoresis, *J. Clin. Microbiol.* **33:**1080–1083.
47. Grattard, F., Pozzetto, B., Tabard, L., Petit, M., Ros, A., and Gaudin, O., 1995, Characterization of nosocomial strains of *Enterobacter aerogenes* by arbitrarily primed PCR analysis and ribotyping, *Infect. Control Hosp. Epidemiol.* **16:**224–230.

48. Kuhn, I., Burman, L. G., Haeggman, S., Tullus, K., and Murray, B. E., 1995, Biochemical fingerprinting compared with ribotyping and pulsed-field gel electrophoresis of DNA for epidemiological typing of enterococci, *J. Clin. Microbiol.* **33:**2812–2817.
49. Valsangiacomo, C., Baggi, F., Gaia, V., Balmelli, T., Peduzzi, R., and Piffaretti, J. C., 1995, Use of amplified fragment length polymorphism in molecular typing of *Legionella pneumophila* and application to epidemiology studies, *J. Clin. Microbiol.* **33:**1716–1719.
50. Thong, K., Ngeow, Y., Altwegg, M., Navaratnam, P., and Pang, T., 1995, Molecular analysis of *Salmonella enteritidis* by pulse-field gel electrophoresis and ribotyping, *J. Clin. Microbiol.* **33:**1070–1074.
51. Millemann, Y., Lesage, M. C., Chaslus-Dancla, E., and Lafont, J. P., 1995, Value of plasmid profiling, ribotyping, and detection of IS200 for tracing avian isolates of *Salmonella typhimurium* and *S. enteritidis*, *J. Clin. Microbiol.* **33:**173–179.
52. Chetoui, H., Delhalle, E., Osterrieth, P., and Rousseaux, D., 1995, Ribotyping for use in studying molecular epidemiology of *Serratia marcescens:* Comparison with biotyping, *J. Clin. Microbiol.* **33:**2637–2642.

12

Utility of PCR *in Situ* for Detecting Viral Infections

GERARD J. NUOVO

1. INTRODUCTION

In situ hybridization is well suited for detecting viral infections. The DNA or RNA one is attempting to localize will have minimal homology with the host's nucleic acid. Further, the sequences of most viral genomes have been elucidated. Indeed, many viral probes are readily available from commercial sources. The acquired immunodeficiency syndrome (AIDS) has underscored the utility of *in situ* hybridization to the diagnostic pathologist, given the high incidence of the many different viral diseases that can occur in a patient with this disease. The one drawback of standard *in situ* hybridization is its relatively high detection threshold. If an infected cell contains 10 or less copies of the viral genome, it is likely to be scored incorrectly as "viral negative" by the *in situ* hybridization assay.[1–3]

Ever since its discovery in 1985, several groups have tried to combine the high sensitivity of the polymerase chain reaction (PCR) with the cell localizing ability of *in situ* hybridization. In 1990 Haase *et al.*[4] was the first to publish data using PCR *in situ* hybridization in a peer-reviewed paper. They employed cell suspensions that, after PCR, were placed on a glass slide which was then subjected to *in situ* hybridization. Although an important beginning, the full utility of *in situ* PCR would require a protocol that could be applied to archival paraffin-embedded, formalin-fixed material. This was first accomplished in a peer-reviewed format in 1991 by Nuovo *et al.*[2] Since that time, over 150 publications have appeared that use *in situ* PCR in their analyses (most listed in Ref. 1).

GERARD J. NUOVO • MGN Medical Research Laboratory, Setauket, New York 11733.

Rapid Detection of Infectious Agents, edited by Specter *et al.* Plenum Press, New York, 1998.

Certainly, the most common application of *in situ* PCR has been in studying the pathogenesis of the human immunodeficiency virus type 1 (HIV-1) infection, and this chapter focuses a great deal on this important issue. Reverse transcriptase (RT) *in situ* PCR has also been extensively used for studying the pathogenesis of RNA viral infections and for understanding the interplay between viruses and host expression in the disease process.[1,5]

2. THE KEY PREPARATORY STEPS

To obtain reproducible and reliable results with *in situ* PCR, one must be aware of certain variables. Before discussing this in the format of the key preparatory steps, the following is a definition of terms used in this chapter.

2.1. Definition of Terms

- PCR *in situ:* A general term for the amplification of DNA or cDNA inside an intact cell.
- *in situ* PCR: Direct incorporation of the labeled nucleotide (or less commonly, primer), into the DNA that is being synthesized in the intact cell during PCR.
- RT *in situ* PCR: Direct incorporation of the labeled nucleotide into the PCR-amplified cDNA in the intact cell.
- PCR *in situ* hybridization: Detection of the PCR-amplified DNA in an intact cell by a hybridization step after PCR by using a labeled probe.

2.2. The Fixative

The most common fixative used by the surgical pathology laboratory is 10% neutral buffered formalin. Formalin fixes tissue by cross-linking the amino groups of proteins and nucleic acids and, in this manner, renders degradative enzymes inactive.[1] As a result of this cross-fixation, the tissue becomes firm, which aids the embedding and sectioning process. After fixation, typically, for 4 to 24 hr for biopsy specimens and 24 hr to several weeks for autopsy material, the tissue is treated with ethanol and xylene so that it can be embedded in melted paraffin at 65 °C. The paraffin solidifies at room temperature, and allows one to cut many thin (usually 4 μM) sections onto a glass side. The paraffin blocks and the slides can be stored for years at room temperature and still permit successful *in situ* hybridization and PCR *in situ*.

Occasionally, other fixatives are used in the surgical pathology laboratory. This usually involves adding an ingredient, such as picric acid (B5 or Bouin's solution, for example) or a heavy metal (Zenker's fixative) to the formalin which allows better preservation of nuclear detail, although this is a somewhat subjective finding. This has important implications for molecular analyses because prolonged fixation ($\geq$8 hr) with picric acid or a heavy metal destroys the signal for *in situ* hybridization or PCR *in situ* hybridization and for solution-phase PCR.[6,7]

Occasionally, the pathologist uses a denaturing fixative, such as ethanol or acetone, or no fixative at all, but rather prepares samples by sectioning at −20 °C with a device called a cryostat. These are often called frozen sections. These maneuvers are usually done to preserve antigens for immunohistochemistry.

We have studied the effect of different fixatives on the end result with PCR *in situ*. One can use either cross-linking or denaturing fixatives. Frozen sections are not recommended. Cross-linking fixatives necessitate a protease digestion step. However, they allow excellent morphology, optimal detection of low copy targets, and, of course, the study of archival material. Denaturing fixatives have the advantage of requiring either no or minimal protease digestion. In my experience, when comparing identical specimens, they do not allow detecting the target sequence in all cells known to contain it, presumably because the amplicon migrates out of the cell which, under optimal conditions, is not detectable.[1,8] However, recently we have been able to eliminate this problem by using a low concentration of proteinase K (from 1 to 10 μg/ml) at room temperature for 15 min (Nuovo, G. J., unpublished observations).

If one can control the fixation process, my recommendation is to use 10% buffered formalin for 1–3 days. Shorter fixation times (8–24 hr) are acceptable for PCR *in situ* hybridization. However, I have found that the longer times (2–3 days) are preferable when doing RT *in situ* PCR, as is discussed shortly.

2.3. The Glass Slide

As someone who was doing *in situ* hybridization in the 1980s, I can attest to the importance of the discovery of silane as a pretreatment step for the glass slide. Before silane, most groups used poly-L-lysine to coat slides to improve sample adhesion to the slide.[9] In my experience, in about 30% of cases, the tissue sections fell off the slides. This was more likely with RNA *in situ* hybridization, where prolonged, high stringency washes were employed. Organosilane pretreatment improved tissue adhesion in our experience to more than 98%.[1] Indeed, if the sections fall off with silane pretreatment, one should suspect that either nonsilane-coated slides were used or water bubbles were trapped under the tissue sections as they were placed on the slides. There are many commercial sources for silane-coated slides. We use slides available from ONCOR (Gaithersburg, MD) and Enzo Labs (Farmingdale, NY).

2.4. The Protease Digestion Step

The most common fixatives in surgical pathology exert their effect by cross-linking the amino groups of proteins to other proteins or to nucleic acids (e.g., 10% buffered formalin) and thus create a complex, three-dimensional network of proteins, DNA, and RNA in the cell. Concomitants of the cross-linking fixative include a rigid, fixed network of charged side chains of the amino acids and pores or channels which, by their size, limit the movement of molecules. This network impedes the entry of reagents, such as probes and Taq polymerases (each about 100 angstroms in size), by physically restricting the movement of such molecules or by limiting the movement of charged molecules through the network of immobile, fixed charged side chains.[1]

If one uses a cross-linking fixative, it is necessary to remove a proportion of the protein–protein and protein–nucleic acid cross-links to allow adequate entry of the larger reagents (e.g., the probe and Taq polymerase) during the hybridization and amplification steps, respectively. One needs to achieve a balance whereby sufficient cross-links between macromolecules are disrupted so that the reagents readily access the target but still preserve morphology. Further, with RT *in situ* PCR, one more extraordinary requirement must be met. The entire genomic DNA must be rendered accessible to DNase digestion before the RT step, or primer-independent DNA synthesis produces a false positive signal during the PCR step.[1,10]

There are many proteases that one can use for PCR *in situ*. The list includes, but is not limited to, pepsin, pepsinogen, trypsin, trypsinogen, proteinase K, and pronase. My recommendation is to use one protease exclusively so to be familiar with its nuances. I prefer pepsin (or trypsin) because it is easily inactivated by rapid washing and increasing the pH of the solution from 2 to 7.5. Further, pepsin is less likely to overdigest the tissue compared with proteinase K.[1] In the concentrations that I typically use (proteinase K, 1 mg/mL), however, an advantage of proteinase K is that it is much less likely to suboptimally digest the tissues which would lead to a false positive result with RT *in situ* PCR. Indeed, suboptimal protease digestion is the most common reason for an unacceptable result from RT *in situ* PCR, as is discussed shortly.

2.5. The Hot Start Maneuver

A common misconception about solution-phase PCR is that it can routinely detect 1–10 copies per sample. This is true if no nontarget DNA is included in the reaction mixture. Of course, in most experiments, one is attempting to amplify a target from a cellular or tissue sample where the nontarget DNA far exceeds the amount of target DNA, typically on the order of millions to billions more sequences of equivalent size. The background DNA competes for the primer. This is called mispriming. Further, the primers, also present at a high concentration, bind to each other. This is called primer oligomerization. In either case, the nonspecific primer binding initiates DNA synthesis during PCR. The end result is to generate nonspecific smears or bands of DNA and, importantly, to reduce the amount of target-specific DNA synthesized. Thus, both the specificity and sensitivity of PCR are compromised.[3,11]

The key to inhibiting the unwanted pathways of mispriming and primer oligomerization is to realize that neither hybridization process will contain much homology. Alternatively, the primer–target hybridized complex should possess 100% homology. Thus, if one withholds a key reagent, such as the Taq polymerase, until the temperature of the amplifying solution is at least 55 °C, then one inhibits the unwanted DNA synthesis pathways yet still allow primer–target annealing to persist. This is the basis of the hot start maneuver. For solution-phase PCR, the hot start maneuver allows reproducibly detecting 1–10 copies with a background of 1 μg of nontarget DNA and a concomitant and striking inhibition of primer oligomerization and mispriming.[3,11] For PCR *in situ* hybridization, the hot start modification allows routinely detecting 1 target per cell using a single primer pair. It allows for target-specific direct incorporation of a reporter nucleotide (e.g., digoxigenin dUTP) for DNA targets if one is using frozen, fixed tissues.[1,3] This is NOT true for paraffin-embedded (or any other heated tissue) because of DNA repair, as is described shortly. Finally, the hot start maneuver is not needed for RT *in situ* PCR because pretreatment with DNase eliminates each of the nonspecific DNA synthesis pathways.[1,8]

There are several ways to achieve hot start PCR. A simple, straightforward way is to withhold a key reagent, such as the $MgCl_2$ or the Taq polymerase, until the temperature of the PCR block reaches at least 55 °C. This so-called manual hot start is done either in solution phase or *in situ*.[1,3] In solution phase, another way to do manual hot start is to separate the amplifying solution from a key reagent with a wax that melts at 55 °C and thus allows all the reagents to mix.[11] This product is sold commercially as Ampliwax™ (Perkin Elmer). Another way to do hot start *in situ* PCR is to add a chemical which either inhibits nonspecific binding of the primer at temperatures below 55 °C or does not allow the Taq polymerase to function below 55 °C. This is chemical hot start and is achieved, for example, by the proper concentration of single-stranded binding protein or by anti-Taq polymerase antibody which dissociates from the polymerase above 55 °C, respectively.[1,8,10] The latter is commercially available as AmpliGold™ (Perkin Elmer). Another way to do chemical hot start is to substitute deoxy uridine triphosphate (dUTP) for deoxy thymidine triphosphate (dTTP) and include the enzyme N-glycosylase in the amplifying solution. This enzyme degrades the DNA synthesized at room temperature and becomes inactive at >55 °C.[1,8,10] This system also has the advantage of degrading the target specific amplicon inadvertently carried over one amplification reaction and thus reduces the risk of carry-over contamination.

2.6. DNase Digestion

The hot start maneuver is needed only for PCR *in situ* hybridization to detect low copy DNA targets. DNase digestion is, clearly, required only for RT *in situ* PCR. After optimal protease digestion, DNase digestion completely eliminates DNA repair, mispriming, and primer oligomerization inside the cell. In turn, this allows directly incorporating the labeled nucleotide into the PCR-amplified target-specific cDNA without the hot start maneuver. This obviates a hybridization step and makes RT *in situ* PCR, in this sense, technically easier and quicker to perform than PCR *in situ* hybridization.

The key for successful DNase digestion is to realize that there is a strong correlation between the length of fixation time and the protease digestion time.[1,8,10] The longer a tissue has been fixed in a cross-linking fixative, such as 10% buffered formalin, the more time it must be digested in protease to allow successful DNase digestion. As an example, skin fixed for 6 hr typically needs 10 min of protease digestion (pepsin, 2 mg/mL), whereas the same tissue fixed for 48 hr may need 90 min of protease digestion. The theoretical basis of this must, in part, reflect the increased protein-DNA cross-links with longer fixation time which must be removed for the DNase to access the entire genomic DNA. Whatever the explanation, the most important concept for RT *in situ* PCR is to realize that the demonstration of optimal protease digestion time REQUIRES certain controls and, without this demonstration, one cannot rely on the results of the assay. The definitions follow:

Optimal protease digestion

• No DNase (positive control):	Intense nuclear signal in all cell types
• DNase, no RT* (negative control):	No signal in any cell type
• DNase, RT with target specific primers (test):	Signal, when present, usually cytoplasmic and restricted to certain cell types

Suboptimal protease digestion (Fig. 1)

• No DNase (positive control):	± nuclear signal in all cell types
• DNase, no RT (negative control):	Nuclear signal (often stronger) in all cell types
• DNase, RT with target specific primers (test):	Nuclear signal in all cell types

Redo experiment increasing the protease digestion time.

*Either omit primers or use irrelevant primers that do not have any corresponding target in the cells being analyzed (Fig. 1).

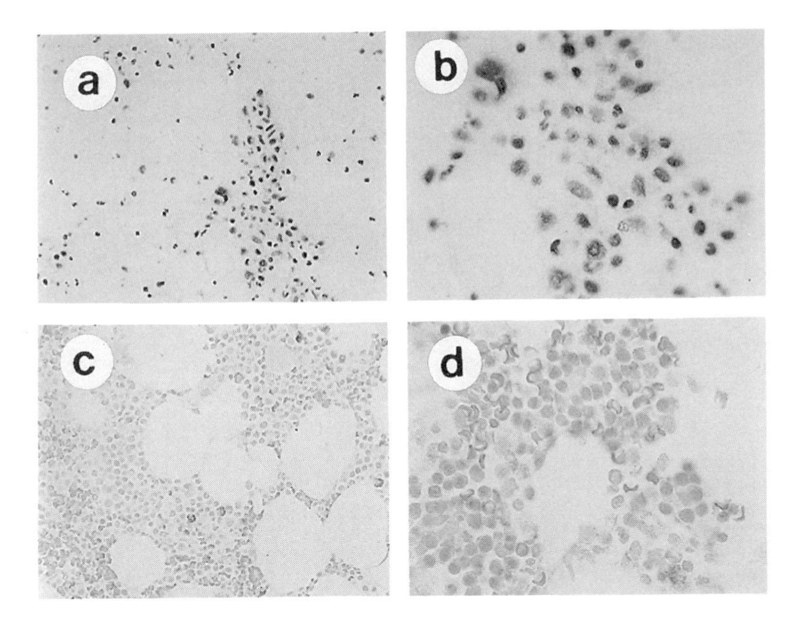

FIGURE 1. The importance of optimal protease time for successful RT *in situ* PCR. This bone marrow biopsy was fixed overnight in 10% buffered formalin. With a protease digestion time of 30 min, a signal is evident with the negative control (a) DNase, no RT and (b) at higher magnification. Note the nuclear localization of the signal and its presence in all the cell types in this tissue. These are useful indicators of a false positive result with RT *in situ* PCR. The protease digestion time was increased to 60 min. Note (c) the loss of the signal with the negative control. A strong signal was seen with the positive control (no DNase, not shown). (D) The test for parvovirus RNA was negative.

The importance of the positive and negative controls cannot be overstated. Note that the positive control is a general control for the PCR reaction and subsequent detection which determines if the reaction conditions are optimal as defined by an intense *nonspecific* signal. A separate and distinct positive control is to use a cell line or tissue known to contain the target mRNA. This also is recommended. Similarly, the negative control as defined is a general control for determining if the DNA synthesis pathways are rendered inoperative by the protease and DNase digestion. If not, then a false positive signal, albeit nuclear-based, follows. A separate and distinct negative control is to use a cell line or tissue known not to contain the target mRNA and to perform RT *in situ* PCR with the target primers. This too is recommended. However, one must always perform the positive (no DNase) and negative (DNase, no RT) controls for every sample at the same time the test is done to assure that there is no residual DNA synthesis (as compared to cDNA amplification) in the cells being tested.

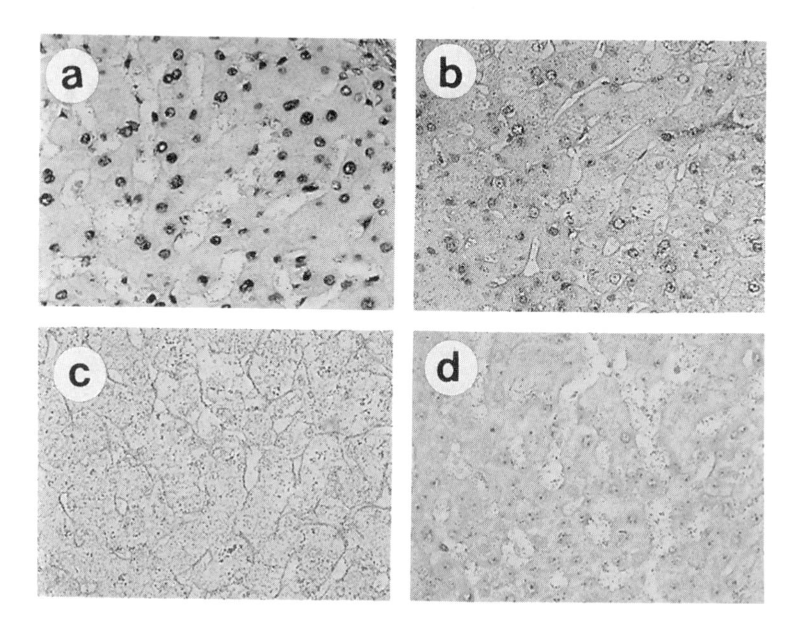

FIGURE 2. Protease overdigestion with RT *in situ* PCR. After 30 min of protease digestion, (a) a strong signal was seen in this liver biopsy with the positive control. (b) After 60 min of digestion, the signal was weaker, localized to the nuclear membrane, and the nuclear detail was less distinct. (c) The signal and the nuclear detail were lost after 90 min of protease digestion. (d) Note the good nuclear detail and lack of signal with the negative control (DNase, no RT) after 60 min of protease digestion.

Protease overdigestion (Fig. 2)

• No DNase (positive control:	Weak cytoplasmic signal, poor morphology
• Dnase, no RT* (negative control)	Weak signal, poor morphology
• DNase, RT with target specific primers (test):	Weak signal, poor morphology

Redo experiment decreasing the protease digestion time.

I am often asked if increasing the DNase digestion time or concentration helps eliminate the background from the negative control. In my experience, this is not the case. When one sees a signal with the negative control (assuming overnight DNase digestion with 10U DNase per tissue section), then one should increase the protease digestion time to eliminate the background. Using an optimal protease digestion, it was noted that 7 hr of DNase digestion is sufficient to eliminate each of the unwanted DNA synthesis pathways.[1]

2.7. Determination of the Optimal Protease Digestion Time for RT *in Situ* PCR

The optimal protease digestion time for PCR *in situ* hybridization (or standard *in situ* hybridization) for most tissues is from 15–30 min (pepsin, 2 mg/mL). This narrow range of optimal protease digestion time for a wide range of tissue types and fixation times (with a cross-linking fixative) probably shows that the putative function of a protease with PCR *in situ* hybridization is simply to allow access of the larger reagents (such as probe and Taq polymerase) to the target. With RT *in situ* PCR, one also must expose the entire genomic DNA so that it is digested by DNase. The simplest and recommended way to determine the optimal protease time for RT *in situ* PCR is to use the so-called start up protocol (Fig. 2). Specifically, place three tissue sections/cell preparations on the slide and protease them for 30, 60, and 90 min, respectively. Then do *in situ* PCR using any primer pair. Because the sample has been heated (either because it was embedded in paraffin or because it was exposed to dry heat at 60 °C for 15 min prior to PCR), then the primer-independent DNA synthesis pathway must be operative, as described later. The optimal protease time is evident as an intense nuclear signal in the majority of the cells.

2.7.1. *Unwanted DNA Synthesis during in Situ PCR*

There are four DNA synthesis pathways which may be operative during *in situ* PCR:

1. DNA repair (primer-independent DNA synthesis);
2. mispriming (primer and nontarget DNA);
3. primer oligomerization (primer and primer); and
4. target-specific amplification (primer and target).

To understand the basis of the DNA repair pathway, a brief review of the normal tissue processing procedure is in order. After fixation with 10% buffered formalin, one must embed the tissue in paraffin wax to be able to cut very thin sections (4 μM) with a knife, called a microtome. To embed the tissue in paraffin, it is first dehydrated in graded ethanol, then placed in xylene. The embedding in paraffin is done at 65 °C for 4 hr. Then it is brought to room temperature where it hardens and can be stored indefinitely.

To study the primer-independent DNA synthesis pathway, one can take expose frozen, fixed tissue to ethanol and xylene or 65 °C heat and then do *in situ* PCR without primers. These types of experiments reveal that the primer-independent signal is operative only if the tissue is exposed to dry heat (≥55 °C) before PCR.[1,10] This is why the primer-independent signal is always evident with paraffin-embedded tissues.

The use of frozen, fixed tissue that has not been exposed to dry heat allows one to study the role of mispriming during *in situ* PCR. If one uses a lymph node for such an experiment and human papillomavirus (HPV)-specific primers, then any signal has to be nonspecific because HPV does not infect lymphoid tissue. Such experiments have shown that no signal is evident if the hot start maneuver is used. However, a mispriming-based signal is evident if all reagents are added at room temperature.[1,10]

Primer oligomerization can be studied by doing *in situ* PCR on paraffin-embedded tissue that has been optimally treated with protease and digested with DNase using a primer pair that does not have a target in the tissue sample. Such experiments have shown that primer oligomerization does not cause a detectable signal inside the cell.[1,10] It is important to realize that primer oligomerization in *the amplifying solution* can cause background during RT *in situ* PCR because the primer oligomer-based DNA synthesis in the amplifying solution incorporates the labeled nucleotide and can stick nonspecifically to cellular proteins and, less likely, to nucleic acids. This problem is corrected by doing a high-stringency wash (60 °C in 15 mM salt for 10 min) after the PCR step.[1]

3. PROTOCOLS

This next section lists step-by-step protocols for PCR *in situ* hybridization and RT *in situ* PCR, respectively.

3.1. PCR *in Situ* Hybridization (from Ref. 1)

1. Place three tissue sections (where possible) on a silane-coated slide.
2. Bake at 60 °C for 15–60 min.
3. Wash in fresh xylene for 5 min, 100% ethanol for 5 min, and air dry.
4. Protease digest in pepsin (2 mg + 9.5 mL water + 0.5 mL 2N HCl) for 30 min, RT.
5. Rinse in water, 100% ethanol, and air dry.
6. PCR step

   Prepare the following solution:

   5 μL of GeneAmp buffer from the Perkin Elmer kit
   9 μL of $MgCl_2$ (25 mM stock)
   8 μL dNTP (final concentration 200 μM)
   1.5 μL 2% BSA (bovine serum albumin)
   3 μL of primer pair (20 μM stock)
   22.5 μL sterile water
   1 μL Taq polymerase (5U/μL) (add at 55 °C, hot start maneuver).*

*One can use several containment systems for PCR *in situ*, including polypropylene coverslips anchored with two drops of nail polish, the Amplicover system from Perkin Elmer, or the Self Seal system from MJ Research Laboratory.

7. Denature at 94 °C for 3 min, then 55 °C for 2 min and 94 °C for 1 min; 35 cycles.
8. 100% ethanol, air dry, and codenature at 95 °C for 5 min with probe cocktail:
   50 μL formamide*
   30 μL 25% dextran sulfate
   10 μL 20X SSC
   10 μL probe (50 to 500 ng/mL)
9. Hybridize for 2 hr at 37 °C.
10. Wash at 60 °C for 10 min in 0.2X sodium chloride citrate (SSC) with 0.2% BSA.
11. Detect labeled probe using appropriate alkaline phosphatase conjugate.
12. Chromogen—nitroblue tetrazolium/bronio-chloro-indolyl-phosphate (NBT/BCIP) for 30–60 min.
13. Nuclear fast red counterstain for 3 min, ethanol, xylene, permount, coverslip.

3.2. RT *in Situ* PCR (from Ref. 1)

1. Place three tissue sections (where possible) on a silane-coated slide.
2. Bake at 60 °C for 15–60 min.
3. Wash in fresh xylene for 5 min, 100% ethanol for 5 min, and air dry.
4. Protease digest in pepsin (2 mg + 9.5 mL water + 0.5 mL 2N HCl) for optimal time.
5. Rinse in RNase free water, 100% ethanol, and air dry.
6. DNase digest two of the three sections overnight at 37 °C using 1 μL RNase-free DNase (Boehringer Mannheim, 10U/μL) + 8 μL diethylcarbonate (DEPC) water + 1 μL PCR buffer II (Perkin Elmer).
7. Remove DNase with RNase-free water, 100% ethanol, and air dry.
8. RT/PCR steps
   Prepare the following solution:
   10 μL EZ buffer (Perkin Elmer RT PCR kit)
   1.6 μL each of deoxy adenosine triphosphate (dATD), deoxy cytosine triphosphate (dCTP), deoxy guanine triphosphate (dGTP), and deoxy thymidine (dTTP) (final concentration 200 μM)
   1.6 μL of 2% BSA
   1.0 μL of RNasin

*The probe cocktail is for full length probes, not oligoprobes. For the latter, I recommend using 10% formamide and changing the posthybridization wash to 50 °C for 10 min in 1X SSC.

3.0 μL of primers 1 and 2 (20 μM stock)*
0.6 μL digoxigenin dUTP (1 mM stock)
14.6 μL RNase-free water
12.4 μL 10 mM $MnCl_2$ or Mn acetate
2.0 μL rTth (2.5 U/μL)

9. RT step—65 °C for 30 min.
10. Denature at 94 °C for 3 min, then 60 °C for 2 min and 94 °C for 1 min; 20 cycles.
11. Depending on containment system, wash in xylene, 100% ethanol, and air dry.
12. Wash at 60 °C for 10 min in 0.2X SSC with 0.2% BSA
13. Detect labeled cDNA using appropriate alkaline phosphatase conjugate.
14. Chromogen—NBT/BCIP for 10–20 min.
15. Nuclear fast red counterstain for 3 min, ethanol, xylene, permount, coverslip.

3.3. Troubleshooting Guide

With respect to PCR *in situ* hybridization, the potential problems are grouped into two categories: no signal when it is expected that the target is present and a signal in most cell types. The former problem should first be addressed by determining if indeed the target sequence is present in the sample of interest. This is done by solution-phase PCR. I have had people send me cervical tissues thought by histology to indicate HPV infection, where on reexamination the histologic features are equivocal for HPV. In such instances, HPV is rarely found by PCR *in situ* hybridization. The absence of the target is confirmed by solution-phase PCR. When no signal is evident but the target is known to be present, increasing the concentration of the probe 10-fold and decreasing the stringency by 15 °C are recommended. If a signal is still not evident, test the probe for incorporation of the labeled nucleotide. If not adequate, the probe should be resynthesized.

In the latter scenario, a signal is sometimes evident in all cell types, including those known not to contain the target sequence. In this instance, a tissue or cell line known not to have the target sequence should also be tested. If it likewise yields a signal, then this is classic background. To resolve this problem, decrease the probe concentration by 10-fold and retest. If background persists, then increase the stringency by 15 °C most simply by increasing the temperature of the posthybridization wash. Another tip is to decrease the hybridization time to 2 hr. With these modifications, the background is eliminated, which is confirmed by using a tissue known not to contain the target sequence.

*For the negative control, use nonspecific primers or omit the primers.

With respect to RT *in situ* PCR, the best troubleshooting advice is described in detail in section 2.6. The most common problem I have seen with RT *in situ* PCR is protease under digestion, recognized as a nuclear-based signal in different cell types in the DNase, no RT control. The experiment must be done by increasing the digestion time. Another common problem is the lack of signal (often with high background) with poor morphology. In this case, decrease the protease digestion time. If no signal is evident and the morphology is good, I recommend the "start-up" protocol, as detailed in Ref. 1, where one takes advantage of DNA repair to generate a signal using *in situ* PCR.

4. HIV-1 as a Model for PCR *in Situ* Hybridization and RT *in Situ* PCR

An important impetus for the development of PCR *in situ* has been understanding the pathogenesis of HIV-1. HIV-1, a retrovirus, synthesizes cDNA from its genomic RNA. Then one copy of the viral DNA integrates into the host genome. The viral DNA can remain quiescent for years. This is the basis of latent infection and the very long subclinical period of disease.[1,4,12] The key variable in clinical disease is the shift from latent to productive infection, marked by the synthesis of a variety of spliced RNAs and infectious virions. The one copy of viral DNA is not detectable by standard *in situ* hybridization.[1] However, it can be detected by PCR *in situ* hybridization. The distinction between latent and activated infection is made by doing PCR *in situ* hybridization for the provirus and RT *in situ* PCR for viral transcripts.[1]

We have performed such analyses on a variety of tissues from people with early HIV-1 infection to end-stage disease. The following information has been ascertained:

1. The viral load, defined by the number of infected cells, is greatest in the lymph node. On average, 30% of the CD4 cells in the lymph nodes are infected by HIV-1 in people who are in the early stages of the disease. The ratio of active:latent infection at this stage is 1:20. There are about 60 million CD4 cells in the lymph node. Thus, about 1 billion cells are actively infected and presumably destroyed each day, even in a person who has no symptoms. At end stage, only about 1 billion CD4 cells remain in the lymph nodes.[1,13,14] Virtually all of these cells are actively infected. Thus, it is evident that end-stage AIDS may be defined as the state whereby the pool of uninfected CD4 cells is depleted to near zero and the remaining CD4 cells are actively infected. At the early stage of the disease, many of the cells are uninfected or latently infected and presumably functional. Still, the actively infected pool of about one bil-

lion cells underscores the need for aggressive and early antiretroviral treatment.

2. Latent HIV-1 infection is routinely detectable in the microglial cells of the central nervous system of people with AIDS, especially in the cerebrum. AIDS dementia is marked by activated infection involving the microglia, astrocytes, and neurons. Further, there is a concomitant upregulation of cytokine expression, including tumor necrosis factor, inducible nitric oxide synthetase, and macrophage inflammatory protein 1∂ and β. Cytokine expression predominates in the area of viral infection, but in the cells uninfected by the virus.[5,15]
3. Sexual transmission of HIV-1 is the primary spreading mode of the virus. HIV-1 localizes to the male spermatogonia and their progeny.[1,16] The infection is activated and destroys the spermatogonia, which explains the testicular atrophy typical of end-stage AIDS. Productive infection and death of the spermatogonia allows release of virions in the semen and facilitates transmission during unprotected sex. HIV-1 localizes to the

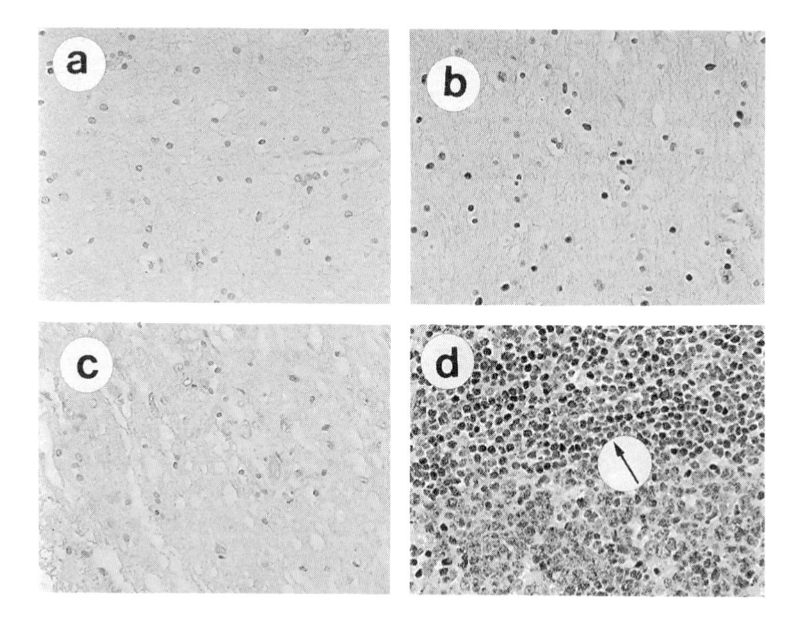

FIGURE 3. (a) HIV-1 DNA was not detected by standard *in situ* hybridization in this CNS tissue from the cerebrum of a patient with severe AIDS dementia. (b) Viral DNA was detected if PCR *in situ* hybridization was done. (c) However, PCR-amplified viral DNA was not detected in the spinal cord, where there was no corresponding symptomatology. (d) Viral DNA was detected in many cells by PCR *in situ* hybridization in the lymph node of a person who had just seroconverted. Many of the infected cells localized to the CD4 rich parafollicular zone.

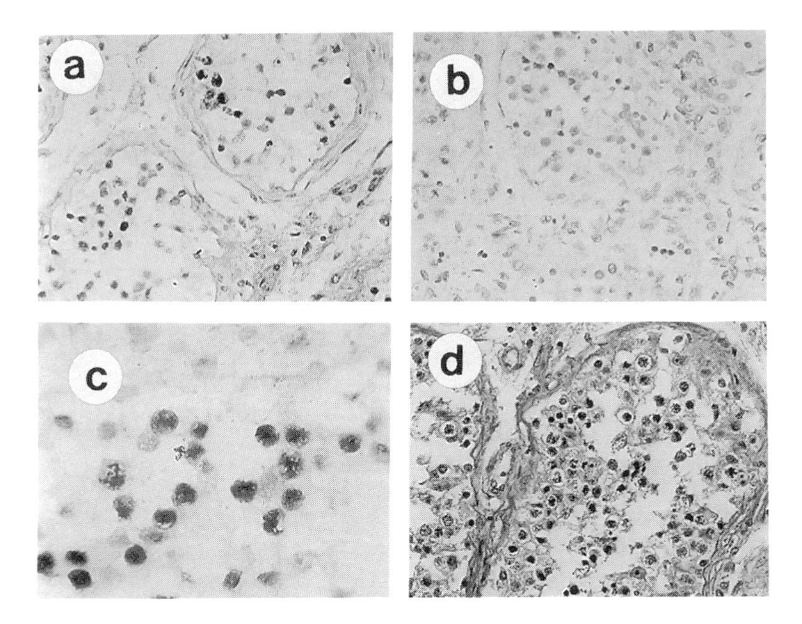

FIGURE 4. Detection of HIV-1 DNA by PCR *in situ* hybridization. (a) PCR-amplified HIV-1 DNA was detected in the spermatogonia and their progeny and (c) at higher magnification in the testes of a man who died of AIDS. (b) Far fewer cells were detected by standard *in situ* hybridization. (D) The corresponding H&E shows the loss of spermatids, typical of end stage AIDS.

endocervical macrophage (and macrophage at the ano-rectal junction) whereby it is transported to the regional lymphatic and thus initiates systemic infection.[1,17] This explains why women are at much greater risk than men for acquiring HIV-1 infection through unprotected sex because the target cell (the endocervical macrophage) is directly exposed to the virus during sex whereas, of course, the spermatogonia is not. The penile urethra has a transformation zone, similar to the cervix and ano-rectal junction, but it contains fewer macrophages than these other sites.

4. AIDS myopathy is associated with an activated infection of the macrophages in the regions of myocyte necrosis.[1,18]

Figures 3 and 4 compare detection of HIV-1 DNA in the CNS, lymph node and genital tract, respectively, using *in situ* hybridization and PCR *in situ* hybridization.

ACKNOWLEDGMENTS. The author greatly appreciates the support of Mr. S. B. Lewis, President of the Lewis Foundation, and Drs. Michael Bukrinsky and Barbara Sherry from the Picower Institute in my research on HIV-1.

REFERENCES

1. Nuovo, G. J., 1996, PCR *in situ* hybridization: Protocols and Applications, 3rd ed., Lippincott-Raven Press, New York.
2. Nuovo, G. J., MacConnell, P., Forde, A., and Delvenne, P., 1991, Detection of human papillomavirus DNA in formalin fixed tissues by *in situ* hybridization after amplification by the polymerase chain reaction, *Am. J. Pathol.* **139:**847.
3. Nuovo, G. J., Gallery, F., MacConnell, P., Becker, J., and Bloch, W., 1991, An improved technique for the detection of DNA by *in situ* hybridization after PCR-amplification, *Am. J. Pathol.* **139:**1239.
4. Haase, A. T., Retzel, E. F., and Staskus, K. A., 1990, Amplification and detection of lentiviral DNA inside cells, *Proc. Nat. Acad. Sci. USA* **87:**4971.
5. Nuovo, G. J. and Alfieri, M. L., 1996, AIDS dementia is associated with massive, activated HIV-1 infection and concomitant expression of several cytokines, *Molecul. Med.* **2:**358.
6. Nuovo, G. J. and Silverstein, S. J., 1988, Comparison of formalin, buffered formalin, and Bouin's fixation on the detection of human papillomavirus DNA extracted from genital lesions, *Lab. Invest.* **59:**720.
7. Greer, C. E., Peterson, S. L., Kiviat, N. B., and Manos, M. M., 1991, PCR amplification from paraffin-embedded tissues: Effects of fixative and fixative times, *Am. J. Clin. Pathol.* **95:**117.
8. Nuovo, G. J., Gallery, F., MacConnell, P., and Bloch, W., 1993, Importance of different variables for optimizing *in situ* detection of PCR-amplified DNA, *PCR Methods Appl.* **2:**305.
9. Crum, C. P., Nuovo, G. J., and Silverstein, S. J., 1988, A comparison of biotin and isotype labelled RNA probes for *in situ* detection of papillomavirus RNA in genital neoplasms, *Lab. Invest.* **58:**354.
10. Nuovo, G. J., Gallery, F., and MacConnell, P., 1994, Analysis of nonspecific DNA synthesis during *in situ* PCR, *PCR Methods Appl.* **4:**342.
11. Chou, Q., Russell, M., Raymond, J., and Bloch, W., 1992, Prevention of pre-PCR mispriming and primer dimerization improves low-copy-number amplifications, *Nucleic Acid Res.* **20:**1717.
12. Bagasra, O., Hauptman, S. P., Lischer, H. W., and Pomerantz, R. J., 1992, Detection of human immunodeficiency virus type 1 provirus in mononuclear cells by *in situ* polymerase chain reaction, *N. Engl. J. Med.* **326:**1385.
13. Embretson, J., Zupancic, M., Ribas, J. L., and Haase, A. T., 1993, Massive covert infection of helper T lymphocytes and macrophages by HIV during the incubation period of AIDS, *Nature* **362:**359.
14. Nuovo, G. J., Becker, J., Fuhrer, J., and Steigbigel, R., 1994, *In situ* detection of PCR-amplified HIV-1 nucleic acids in lymph nodes and peripheral blood in asymptomatic infection and advanced stage AIDS. *J. Acquired Immune Defic. Syndr.* **7:** 916.
15. Nuovo, G. J., Gallery, F., MacConnell, P., and Braun A. 1994, *In situ* detection of PCR-amplified HIV-1 nucleic acids and tumor necrosis factor RNA in the central nervous system, *Am. J. Pathol.* **144:**659.
16. Nuovo, G. J., Becker, J., Simsir, A., and Shevchuck, M., 1994, *In situ* localization of PCR-amplified HIV-1 nucleic acids in the male genital tract, *Am. J. Pathol.* **144:** 1142.
17. Nuovo, G. J., Forde, A., MacConnell, P., and Fahrenwald, R., 1993, *In situ* detection of PCR-amplified HIV-1 nucleic acids and tumor necrosis factor cDNA in cervical tissues, *Am. J. Pathol.* **143:**40.
18. Seidman, R. and Nuovo, G. J., 1994, *In situ* detection of PCR-amplified HIV-1 nucleic acids in skeletal muscle in patients with myopathy, *Mod. Pathol.* **7:**369.

Index

Page numbers followed by a "*t*" or "*f*" represent tables and figures respectively.